Building Pathology and Rehabilitation

Volume 29

This book series addresses the areas of building pathologies and rehabilitation of the constructed heritage, strategies, diagnostic and design methodologies, the appropriately of existing regulations for rehabilitation, energy efficiency, adaptive rehabilitation, rehabilitation technologies and analysis of case studies. The topics of Building Pathology and Rehabilitation include but are not limited to - hygrothermal behaviour - structural pathologies (e.g. stone, wood, mortar, concrete, etc…) - diagnostic techniques - costs of pathology - responsibilities, guarantees and insurance - analysis of case studies - construction code - rehabilitation technologies - architecture and rehabilitation project - materials and their suitability - building performance simulation and energy efficiency - durability and service life.

João M. P. Q. Delgado
Editor

Advanced Materials, Corrosion Science, and Rehabilitation of Concrete Infrastructure

Springer

Editor
João M. P. Q. Delgado
CONSTRUCT, Department of Civil
and Georesources Engineering
Faculty of Engineering of the University
of Porto
Porto, Portugal

ISSN 2194-9832 ISSN 2194-9840 (electronic)
Building Pathology and Rehabilitation
ISBN 978-3-032-28850-9 ISBN 978-3-032-28851-6 (eBook)
https://doi.org/10.1007/978-3-032-28851-6

This Springer imprint is published by the registered company Springer Nature Switzerland AG
The registered company address is: Gewerbestrasse 11, 6330 Cham, Switzerland

Preface

Building rehabilitation and long-term performance of concrete infrastructure are central challenges for modern engineering, demanding a deep understanding of material behaviour, degradation mechanisms, and innovative solutions for structural preservation. As concrete remains the backbone of global construction—valued for its strength, versatility, and durability—its continued performance depends on the ability to diagnose, protect, and enhance reinforced systems exposed to increasingly complex environmental and mechanical demands.

The main purpose of this book, *Advanced Materials, Corrosion Science, and Rehabilitation of Concrete Infrastructure*, is to provide a collection of recent research contributions that explore emerging materials, corrosion processes, structural pathologies, and cutting-edge technologies aimed at improving the resilience of concrete structures. This volume highlights advancements in coated steel reinforcement, the durability of LC3 concrete, corrosion challenges in post-tensioned systems, and the evolving role of GFRP materials in structural applications. It also addresses the growing integration of artificial intelligence as a powerful tool for diagnosing and rehabilitating infrastructures affected by dynamic actions such as vibrations.

Organized into six chapters, the book provides a concise yet comprehensive overview of current knowledge for engineers, researchers, students, and practitioners across civil, structural, mechanical, and materials engineering. By bridging traditional durability science with innovative materials and intelligent assessment techniques, this work contributes to the ongoing development of safer, more sustainable, and longer-lasting concrete infrastructure.

Porto, Portugal
João M. P. Q. Delgado

Contents

Corrosion Resistance and Electrochemical Techniques for Evaluating the Performance of Coated Steel Reinforcements in Concrete

S. R. Aswanth, Chandru Pichaimuthu, and Karthikeyan Manickam

Abstract Corrosion of steel reinforcement remains the primary cause of deterioration in reinforced concrete (RC) structures exposed to aggressive environments. The increasing demand for durable infrastructure and extended service life, has led to the use of coated steel reinforcement as a corrosion mitigation strategy. This chapter presents various coating systems, their corrosion mechanism, durability and the electrochemical performance evaluation of these coated rebars. Particular emphasis is placed on the barrier effects, sacrificial protection mechanism, coating degradation, and sustainability considerations associated with coated steel reinforcements.

Keywords Coated steel rebars · Corrosion mechanism · Assessment techniques

1 Introduction

1.1 Steel Reinforcement in Concrete Structures

Reinforced concrete (RC) is the most widely used construction material worldwide for infrastructure, including buildings, bridges, and marine structures. This is because of the ability of steel to carry the tensile strength of concrete as concrete is strong only in compression. In this manner, the steel and concrete complement each

S. R. Aswanth · C. Pichaimuthu (✉)
Department of Civil Engineering, National Institute of Technology Calicut, Calicut, Kerala, India
e-mail: chandru@nitc.ac.in

S. R. Aswanth
e-mail: aswanth_p250533ce@nitc.ac.in

K. Manickam
Department of Civil Engineering, National Institute of Technology Rourkela, Rourkela, Odisha, India
e-mail: mkarthi@nitrkl.ac.in

João M. P. Q. Delgado (ed.), *Advanced Materials, Corrosion Science, and Rehabilitation of Concrete Infrastructure*, Building Pathology and Rehabilitation 29,
https://doi.org/10.1007/978-3-032-28851-6_1

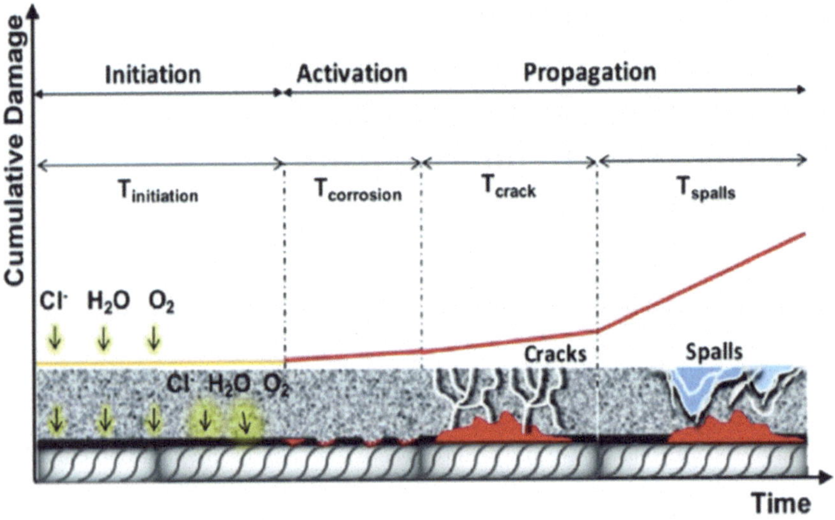

Fig. 1 Initiation and propagation of corrosion of steel in RC (Tutti 1982)

other in a reinforced concrete system. Therefore, providing steel reinforcement in concrete improves the load-bearing capacity, crack resistance and overall structural performance of RC members (Fig. 1).

The use of steel reinforcement in concrete began in the late nineteenth century, when engineers recognised that embedding steel rebars could compensate for concrete's low tensile strength. The steel is protected by a passive oxide film formed on the surface due to the highly alkaline environment of the concrete. Corrosion initiates once this passive film is broken down by the ingress of chlorides, carbon dioxide that is penetrated to the concrete through the cover block. The electrochemical processes that initiate once the corrosion begins form rust products that would expand and create internal stresses in the concrete, eventually leading to cracking and deterioration (Ali et al. 2024; He et al. 2025).

1.2 *Need for Corrosion Protection*

Corrosion of steel reinforcement is widely recognised as one of the primary causes of deterioration in reinforced concrete structures, which directly affects both durability and structural performance (He et al. 2025). Corrosion leads to premature repair of infrastructure, resulting in higher repair costs and reduced service life of the structure. Corrosion related damage causes a severe financial burden on the national infrastructure systems, thus bringing in the need for effective corrosion mitigation strategies (Shokri 2025) (Fig. 2).

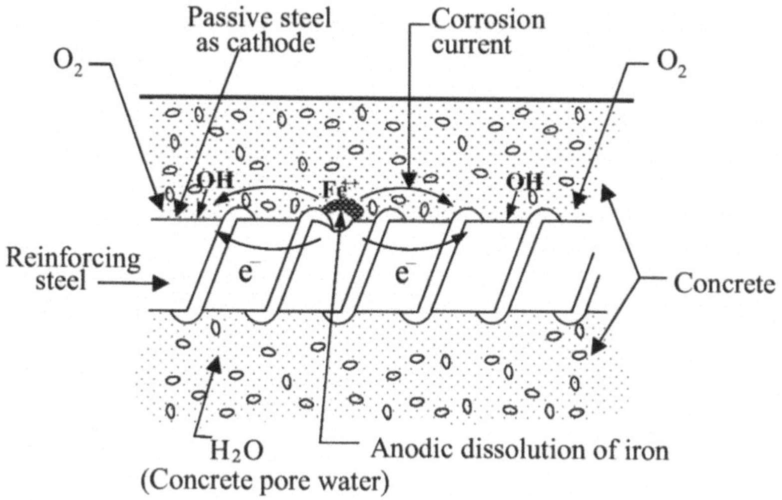

Fig. 2 Schematic representation of the mechanism of corrosion (Ibrahim et al. 2017)

Corrosion also brings in serious safety concerns. The reduction in steel cross-sectional area due to corrosion decreases the load-carrying capacity of reinforced concrete members. In addition, the expansive nature of corrosion products causes cracking and spalling of the concrete cover, further accelerating deterioration and weakening the bond between steel and concrete. Even small defects in structure can cause corrosion and initiate cracking and thus affecting the structural system. This clearly shows the importance of corrosion protection for the safety of structures. (Ali 2025). These effects clearly show that corrosion is not only a durability issue but also a structural reliability problem. Therefore, it is essential to adopt effective corrosion prevention and protection methods to ensure the long term performance and safety of concrete structures.

1.3 Scope and Organization of the Chapter

This chapter evaluates the performance of different types of coated steel reinforcements and understanding their corrosion mechanism. The chapter includes the electrochemical processes causing corrosion initiation and propagation in reinforced concrete. The chapter also says about the need for coating protection and discusses about the different types of metallic and non-metallic coatings such as galvanized, fusion bonded epoxy coating, cement polymer composite coating. The scope also lies on the emerging polymer and nano coatings-based protection strategies. The major emphasis is on understanding how these coatings interact with the concrete environment, under chloride and carbonation-prone exposure conditions.

The chapter further examines the mechanisms by which coatings provide protection, including barrier action, sacrificial (cathodic) protection, and inhibitive or hybrid mechanisms. The limitations of the coating systems, including coating defects, damage during handling and the risk of under-film corrosion, are discussed in this chapter. To evaluate the corrosion resistance and overall performance of coated reinforcements, the chapter will also cover the electrochemical and mechanical assessment methods, including half-cell potential measurements, linear polarization resistance, electrochemical impedance spectroscopy, accelerated corrosion testing and bond strength evaluation. This chapter also says about the changes in service life when coated reinforcements are used and also the sustainability aspects associated with the use of coated steel reinforcements in the RC structures. In overall, the chapter aims to provide an understanding of the corrosion protection provided by the coated steel reinforcement.

2 Fundamentals of Corrosion of Steel in Concrete

The reinforced concrete provides protection to the underlying steel through concrete cover and alkalinity. When the passive state of steel is disrupted, this leads to corrosion initiation and propagation.

2.1 Electrochemical Principles

Corrosion of steel in concrete is fundamentally an electrochemical process involving anodic metal dissolution and cathodic reduction reactions occurring on the steel surface. The reinforced steel rebar behaves as an electrode where localized anodic and cathodic sites are formed due to oxygen availability, moisture content and the chloride content that is present in the concrete matrix. The electrochemical reactions at the anode and cathode are as follows: Eqs. (1) and (2).

At anodic regions, iron dissolves as

$$Fe \rightarrow Fe^{2+} + 2e^{-} \tag{1}$$

At the cathodic sites, oxygen reduction occurs in the alkaline pore solution

$$O_2 + 2H_2O + 4e^{-} \rightarrow 4OH^{-} \tag{2}$$

The electrochemical reactions occur in the concrete pore solution that contains hydroxyl ions, alkali ions, and dissolved oxygen. The electrical circuit is completed by ionic transport in the pore solution and by electronic conduction through the steel reinforcement. The corrosion behaviour is strongly influenced by the moisture content and pore structure of concrete, which control oxygen availability and

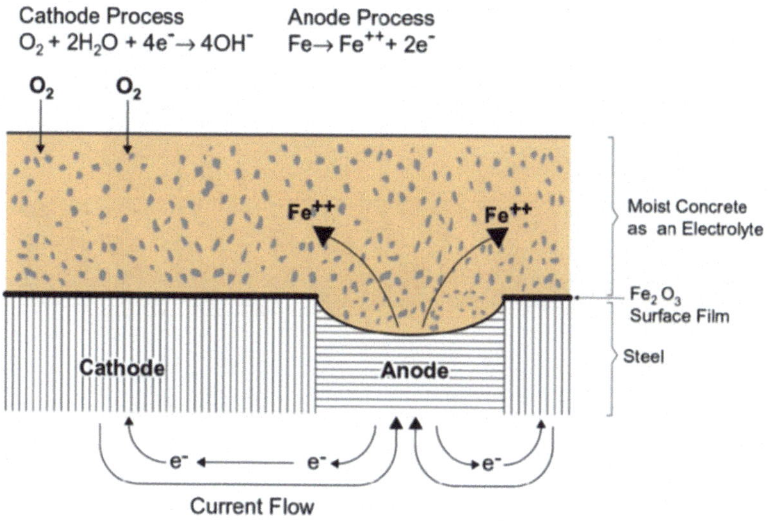

Fig. 3 Electrochemical mechanism of steel corrosion in concrete (Mehta et al. 2006)

ionic conductivity. A denser pore structure increases electrical resistivity and reduces corrosion rates, whereas higher moisture content enhances electrochemical activity and corrosion kinetics (Angst et al. 2020). Macrocell corrosion may also occur due to differences in exposure conditions along the reinforcement length, producing large-scale electrochemical cells that accelerate localized corrosion damage (Jiang et al. 2021) (Fig. 3).

2.2 *Passive Film Formation in Alkaline Concrete*

Reinforcing steel embedded in concrete under normal conditions remains in a passive state due to a highly alkaline pore solution (pH $\approx$ 12.5–13.5). This alkalinity promotes the formation of a thin, protective passive film on the steel surface, which significantly prevents corrosion of reinforcement. The passive film primarily consists of iron oxides and hydroxides, which form a barrier that inhibits further anodic reaction. When the passive film is intact, the corrosion current density is extremely low, often less than 0.1 $\mu A/cm^2$, indicating negligible corrosion activity (Ming et al. 2021), which is illustrated in Fig. 4. The composition and stability of the passive film are influenced by the chemical constituents of the concrete pore solution, particularly calcium and hydroxyl ion concentrations. These components form a more stable, less permeable oxide layer, thereby retarding chloride-induced depassivation (Nong et al. 2025). However, the passive film is thermodynamically stable only within a certain pH range. A reduction in alkalinity due to carbonation or ingress of aggressive ions affects the stability of the film, making steel vulnerable to corrosion initiation.

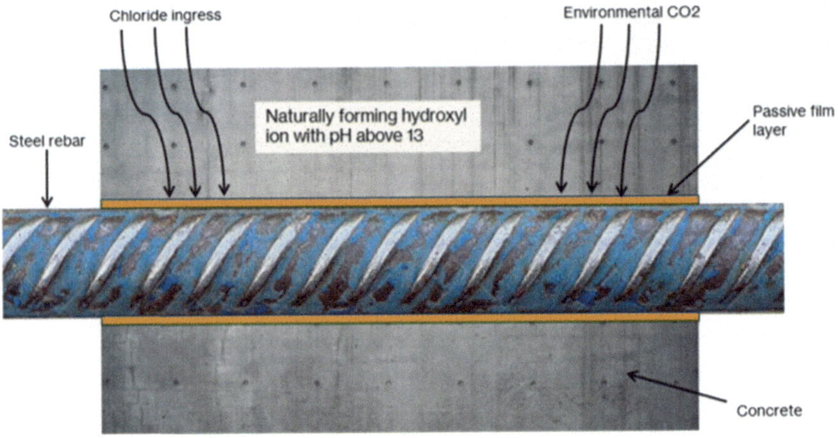

Fig. 4 Conceptual illustration of passive film layer in RC (Osibodu et al. 2025)

2.3 Chloride Induced Corrosion-Mechanism

Chloride ingress is one of the main causes of corrosion initiation in reinforced concrete structures. Chloride ions penetrate the concrete surface through diffusion, capillary action etc., and reaches the steel surface. This penetration is accelerated on the basis of pores structure of concrete, concrete cover and exposure conditions. The corrosion initiation is basically governed by the chloride ingress mechanism described by Fick's second law of diffusion and multi-ionic transport processes at the steel concrete interface (Sripriya et al. 2023) (Fig. 5).

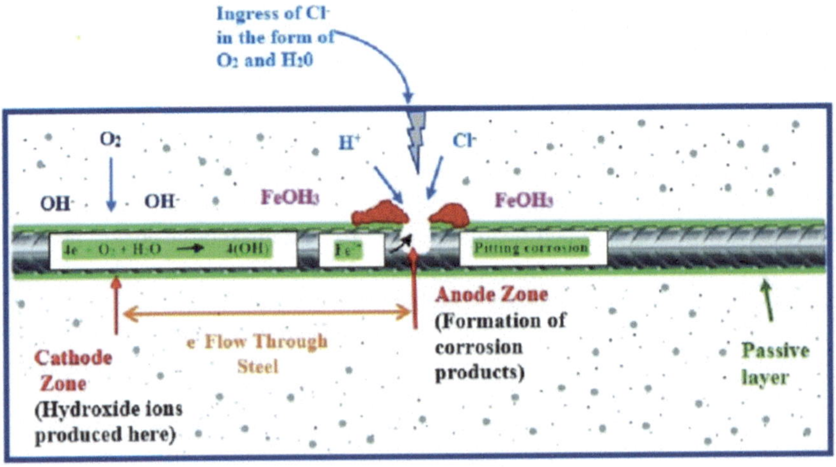

Fig. 5 Mechanism of chloride-induced corrosion (Brindha et al. 2023)

As the free chloride concentration at the steel concrete interface exceeds a critical chloride threshold (Clth) depassivation of the steel rebar occurs at the localised point. The magnitude of Clth is not constant and it depends on the pore solution chemistry, binder composition and pH buffering capacity of concrete. Inclusion of supplementary cementitious materials may modify chloride binding capacity and ionic resistivity, thus influencing the chloride threshold and initiation (Sripriya and Pillai 2021).

After depassivation, corrosion progresses through coupled anodic and cathodic electrochemical reactions. The initial anodic dissolution is given by Eq. (1). Ferrous ions thus produced react with chloride to form complexes as given in Eq. (3).

$$Fe^{2+} + 2Cl^{-} \rightarrow FeCl_2 \tag{3}$$

These complexes further undergo hydrolysis in the alkaline pore solution as given in Eq. (4)

$$FeCl_2 + 2H_2O \rightarrow Fe(OH)_2 + 2HCl \tag{4}$$

The generated HCl causes further lowering of pH and promotes continued dissolution of iron and thereby causing corrosion propagation. The cathodic oxygen reduction occurs as per the Eq. (2). The hydroxyl ions thus produced further reacts with the ferrous ions as in Eq. (5).

$$Fe^{2+} + 2OH^{-} \rightarrow Fe(OH)_2 \tag{5}$$

This ferrous hydroxide further oxidizes to ferric hydroxide as governed by the Eq. (6)

$$4Fe(OH)_2 + O_2 + 2H_2O \rightarrow 4Fe(OH)_3 \tag{6}$$

These corrosion products are further transformed into rust (FeOOH and $Fe_2O_3.H_2O$) and causes tensile stress within the surrounding concrete. This occurs in a cyclic way and thereby making the chloride induced corrosion as autocatalytic and thus causing a sudden propagation once the corrosion is initiated. All the free chlorides that are present in concrete are not causing corrosion initiation. Some of the free chlorides chemically bounds with the hydration products such as Friedel's salt, thus reducing the concentration of available free chlorides. The chloride binding capacity thus depends on the binder chemistry and the presence of fillers (Harilal et al. 2020).

2.4 Carbonation-Induced Depassivation

Carbonation induced corrosion mainly occurs in the structures that are exposed to atmospheric environments and mainly in urban areas. As compared to the chloride induced corrosion, that occurs by pitting, in the case of carbonation induced corrosion leads to the uniform depassivation of steel reinforcement by reducing the alkalinity of the concrete pore solution around the reinforcement. Carbonation is a physiochemical reaction that occurs due to the diffusion of CO_2 in the astmosphere to the pore solution of concrete. This results in the formation of carbonic acid, which is explained by the Eq. (7).

$$CO_2 + H_2O \rightarrow H_2CO_3 \tag{7}$$

The carbonic acid further dissolves to form carbonate ions as shown in Eqs. (8) and (9).

$$H_2CO_3 \rightarrow H^+ + HCO_3^- \tag{8}$$

$$HCO_3^- \rightarrow H^+ + CO_3^{2-} \tag{9}$$

This further reacts with the portlandite to form calcium carbonate as in Eq. (10)

$$Ca(OH)_2 + CO_3^{2-} \rightarrow CaCO_3 + 2OH^- \tag{10}$$

Thus, the carbonation reaction can be summarised as in Eq. 11

$$CO_2 + Ca(OH)_2 \rightarrow CaCO_3 + H_2O \tag{11}$$

Thus carbonation involves the ingress of CO_2 into the concrete pore water, thus forming carbonate ions and the formation of carbonate ions and thus eventually the formation of $CaCO_3$. This results in the further decalcification of CSH gel and calcium hydroxide which are the hydration products (Dhanya et al. 2019) (Fig. 6).

The reinforcing steel in concrete remains passive when the pH of the surrounding concrete is above 12. But, the carbonation of concrete lowers the pH to around 8–9, thus the passive oxide film is disturbed and corrosion is initiated. Thus, carbonation induced corrosion can be considered as a chemical process that occurs due to the reduction in pH rather than due to the ionic attack as in chloride induced corrosion. Carbonation progresses gradually from the surface towards the reinforcement, forming a carbonation depth. The depth of the carbonation is explained on the basis of square root law based on diffusion theory. Thus, excluding the climatic variability carbonation coefficient depends on concrete mix proportion parameters such as binder content, cover depth, water to binder ratio and the time to corrosion initiation is based on this (Sundar and Pillai 2025).

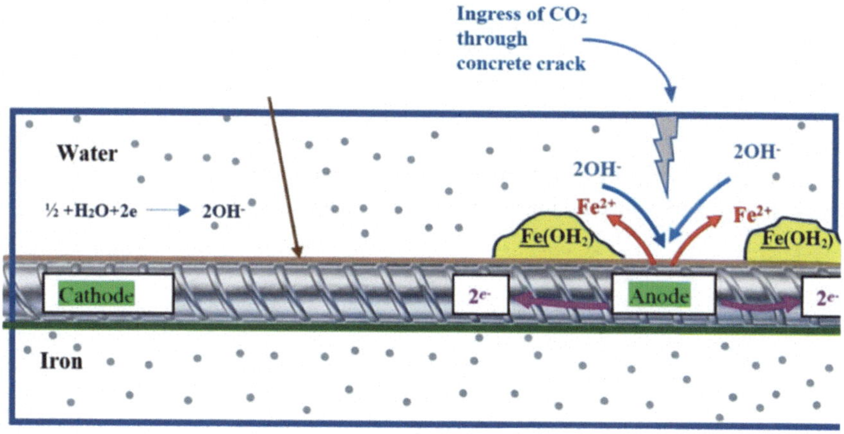

Fig. 6 Mechanism of carbonation induced corrosion (Brindha et al. 2023)

2.5 *Corrosion Propagation and Cracking Mechanism*

As the steel is depassivated by either chloride or carbonation ingress, corrosion propagation begins. This results in the formation of rust, that occupies up to six times volume greater than the original steel volume. This would generate tensile stresses in the surrounding concrete eventually leading to cracking and spalling of the concrete. In chloride induced corrosion, as the passive film is destroyed at the localized region anodic sites are formed on the steel surface. The other areas still remains as passive and acts as cathodes. This leads to severe localized loss of steel cross section even though there is less corrosion rate in overall. Along with these variations in chloride concentration and the availability of oxygen can lead to marcocell corrosion along the length of reinforcement that accelerates the rate of corrosion. As the corrosion progresses, the corrosion products accumulate in the pores near the reinforcement and finally when it exceeds the tensile strength of concrete, radial cracks initiate at the steel concrete interface and eventually propagate outwards.

As discussed earlier unlike chloride induced corrosion, carbonation induced corrosion causes uniform corrosion along the reinforcement length and causes a widespread reduction in steel cross section. Carbonation neutralizes the alkaline pore solution and thus allowing the oxidation reactions to proceed uniformly on exposed steel surface (Montero et al. 2023). Corrosion propagation mainly depends on the environmental humidity and oxygen diffusion in carbonated concrete. In carbonation also corrosion proceeds through th electrochemical reaction in which anodic dissolution and cathodic oxygen reduction reactioons are occurring and corrosion products are formed as in Eq. (6). Though the microstructural changes that are occurring during carbonation densifies the pore structure, but reduces the alkalinity of concrete required for long term protection (Dhanya et al. 2019). In the real structures, both chloride ingress and carbonation occurs simultaneously causing complementary deterioration. Carbonation reduces the alkalinity of concrete and helps in chloride mobility by decomposing binding chlorides, that releases bound chlorides to the pore solution and ease their movement towards the steel. Such kind of coupled mechanism is a critical factor that is influencing corrosion propogation in blended cement systems and varying micro climatic exposure conditions (Das et al. 2025).

3 Coated Reinforcement Systems

Coated reinforcement systems are now emerging as one of the effective protection mechanism against the rebar corrosion. When compared to the conventional protection methods such as increasing the cover thickness and developing mix of higher strength with some supplementary cementitious materials, the use of coated steel reinforcements provides a direct protection on the steel rebar, thereby further delaying corrosion initiation and propagation. The coated reinforcement selected should be based on the specific service environment, expected structural lifespan and cost considerations (Qian et al. 2025).

3.1 Rationale for Coating Protection

Although the concrete provides an inherent protection to the corrosion due to the high pH of the pore solution inside the concrete, the exposure conditions and the defects if any in the concrete eventually leads to the depassivation of the steel reinforcement in the concrete. Thus the protective coatings will act as an additional defensive layer, reducing the probability of the corrosion initiation and thereby extending the service life (Mukhopadhyay et al. 2025). The preliminary purpose of providing coating to the steel reinforcement is to prevent the direct contact between the steel rebar surface and corrosion initiation agents, thereby delaying the onset of corrosion and increasing the service life structure. But the coating protection and performance mainly rely on the coating integrity, adhesion of coating to the steel surface and the resistance against the damages induced during manufacturing and handling. Defects if any such as potholes, holidays and scratches can act as spot of localized corrosion initiation sites and thus significantly affects the long term durability (Qian et al. 2025). Along with this, the variations in the construction quality, cracking and environmental exposure cycle also influences the chloride penetration and carbonation depth, thus reducing the effectiveness of passive film. So, the coating of reinforcement acts as a robust method for achieving the extended service life.

3.2 *Types of Coating*

Coatings are basically classified on the basis of the material composition of the coating system as well the type of protection mechanism that the coating is offering. The major coating systems that are discussed in this chapter are organic coatings, metallic coatings and the emerging type of polymer and nano material based coating.

3.2.1 Fusion Bonded Epoxy Coated Reinforcement

Epoxy coated reinforcement has a steel reinforcement that is coated with a thin layer of epoxy resin which will act as protective interface between the steel surface and the surrounding concrete pore solution. Epoxy coatings are basically thermosetting organic polymers coatings that are composed of epoxy resins, curing agents, pigments and fillers. Epoxy coating are applied using an electrostatic spray process onto a pre-heated reinforcing bars at around 230 °C and cured further to produce a polymeric film on the steel surface. The fusion process will ensure strong adhesion between the coating and the reinforcement surface and thus producing a uniform coating thickness in the range of 175–300 μm (Kamde and Pillai 2020a, b). The typical coating thickness and coating continuity required as per the international standards is given in Table 1.

Epoxy coatings are having high chemical stability in alkaline pore solution. The surface of the epoxy coating appears to be the smooth due to adhesion of epoxy to the steel surface as in Fig. 7. These epoxy coatings maintain stability in high pH environments and effectively reduce the chloride diffusion to the steel surface. The epoxy coated bars have to satisfy all the standards as prescribed in international standards. The fusion bonded epoxy coated bars acts as a good alternative to improve the corrosion resistance and provides a barrier to the attack of corrosive chemicals and makes it suitable to use in extreme environments (Saraogi and Warudkar 2018).

Table 1 Specifications for fusion bonded epoxy coated rebars

Standard	Typical coating thickness	Coating continuity
IS 13620:1993	Atleast 90% of all coating thickness shall be between 0.1 and 0.3 mm after curing	Not more than an average of two holidays per 300 cm of coating
ASTM A775/775 M	Coating thickness for bar size Nos. 3–5 (175–300 μm) and for Nos. 6–18 (175–400 μm)	Not more than 3 holidays per metre
ISO 14654:1999	Coating thickness after curing shall be 175–300 μm and not applies to areas of defects	Not more than 4 holidays per linear metre
BS 7295-1:1990	95% of thickness should be within 150–280 μm	Not more than 5 holidays per linear metre

Fig. 7 Typical microstructure of epoxy coating

During the storage of fusion bonded epoxy coated bars care should be taken to avoid exposure to UV radiation, as well as during the handling and transportation. Prolonged UV exposure can degrade the epoxy coating and reduces the corrosion resistance, thus the exposure to the sunlight should be minimized (Kamde and Pillai 2020a, b). The damages that occur due to the bending, dragging and improper handling in the site can cause corrosion initiation at that localized region even in the absence of extreme exposure conditions. This indicates the importance of quality and standards that has to be followed during the manufacturing stage (Kamde and Pillai 2017). The presence of epoxy further affects the mechanical interlocking between the rebar and the concrete. Thus, they may exhibit a slightly lower bond strength when compared to the normal uncoated bars due to the reduced frictional resistance at the surfac e of the rebars. But in most of the cases, the reduction in bond strength is within the acceptable limits and can be compensated by taking appropriate design considerations.

Field implications world-wide points that the fusion bonded epoxy coated rebars provides improved durability and delayed corrosion initiation when compared to the conventional reinforcement. But there are some performance variations that are observed based on the coatiung thickness, exposure to environment and thus have shown earlier corrosion initiation. These varying performances points that even though the fusion bonded epoxy coated bars has better corrosion resistance, but this depends on the coating process, exposure of bars and the handling practices in construction site. In hot dip galvanizing.

3.2.2 Galvanized Reinforcement

In galvanized reinforcement, the steel reinforcement is coated with a layer of zinc, that is mainly applied through the hot dip galvanizing process. In this type of coating protection to the steel reinforcement, the underlying steel reinforcement is sacrificially protected by the zinc coating. In hot dip galvanizing, the steel reinforcement bar is immersed in the molten zinc. This results in the formation of a bonded coating that has an outer zinc layer and inner Fe-Zn intermetallic alloy layers. Due to thisa metallic bonding, the galvanized bars are having better mechanical properties and improved adhesion when compared to other types of coatings. The typical thickness of galvanized bars are in the range of 80–180 μm (Patnaik et al. 2019). The values obtained from different international standards are given in Table 2

As shown in Fig. 8, the galvanized bars are exhibiting a layered microstructure that is composed of gamma, delta, and zeta layers that exist beneath the outer pure zinc layer. The variations that are seen in the thickness as well as the changes in the composition of different layers, can affect the corrosion performance and bond strength of galvanized bars. Galvanizing of the bars would provide better corrosion resistance but in ternary blended cement it causes higher current densities and higher coating thickness (Molina et al. 2017). When the corrosion resistance of the galvanized bars are compared with the epoxy coated reinforcements under damage condition also shows the better corrosion resistance of the galvanized bars and the cathodic corrosion current density did not rise above 0.35 $\mu A/cm^2$ (Yoon and Nam 2024).

When the galvanized bars are undergoing an exposure with the fresh concrete, the zinc in the coating undergoes reactions that would result in the evolution of hydrogen gas and zinc corrosion products. These reactions further influence the interfacial transition zone (ITZ) between the steel and concrete. This hydrogen evolution can increase the porosity in the ITZ and cause a slight reduction in the bond strength (Pokorny et al. 2024). But there is limited effect on the bond strength in the long term as long as the coating remains intact. Similarly, the galvanized coatings also perform better in the carbonated concrete by delaying the corrosion and makes it useful for structures exposed to urban environments (Wang et al. 2024). Thus, the galvanized bars has better corrosion resistance, improved service life and mechanical and structural performance when compared to the uncoated reinforcement.

Table 2 Specifications for galvanized rebars

Standard	Typical Coating thickness	Coating continuity
IS 19068:2025	Average thickness of zinc alloy coating shall be at least 50 μm	Measured thickness at any location shall be atleast 40 μm
ASTM A767/767 M	Thickness to be more than 86 μm for bar designation No. 3 and larger	Bars should have no uncoated areas and should be free of blisters, flix spots or inclusions
ISO 1461:2009	Mean coating thickness should be atleast 85 μm for bars > 6 mm	Coating thickness at any location should be > 70 μm
EN10348:2024	Mean thickness should be 610 g/m^2 for bars > 6 mm and 505 g/m^2 for bars < 6 mm	Coating should be continous with no bare steel, ribs exposed even after galvanizing

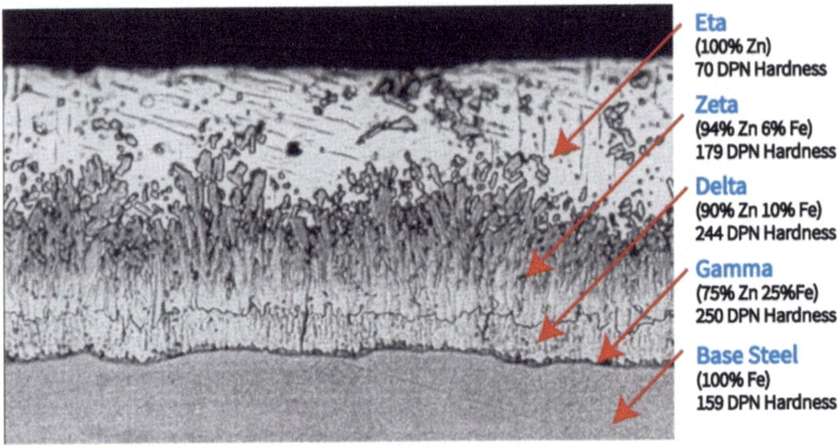

Fig. 8 Composition of hot dipped galvanized bars (American Galvanizer's association)

3.2.3 Cement Polymer Composite Coated Reinforcement

Cement polymer composite coatings (CPCC) unlike the metallic or epoxy coatings, has a hybrid system that is composed of cementitious materials that are blended or mixed with the polymers. CPC coatings have a mixture of polymer emulsion, binder materials basically cement or fly ash and other fillers inorder to improve the adhesion and flexibility of the coating. The polymer phase usually with a solid content of about 45% forms a continuous film over the rebar and the cementitious part eventually hydrates and form the hydration products and thereby gives a dense microstructure to the coating. Thus CPCC is said to have the hybrid composition of both the organic and inorganic coatings. The cement hydration phase thus provides a dense microstructure and compatibility with the surrounding concrete.

The thickness of the coating after the application should be around the range of 175 μm (Kamde and Pillai 2020a, b). There are no specific standards for the CPCC coated bars and is explained by the performance specifications as protective coatings. The recommended thickness based on several literature or guidelines is given in Table 3 above. The cementitious phase forms a good mechanical interaction with the concrete and the polymer phase enhances the adhesion and flexibility of the coating. Also, the polymer modified cementitious coatings have the ability to form a dense interfacial film with the steel and thereby the resistance against the corrosion increases (Kamde and Pillai 2022). The CPCC coatings can also be applied directly at the site when compared to the other metallic and non-metallic coatings that requires specific factory conditions for the applications. This improves their large scale on site applicability and popularity.

Figure 9 shows the microstructure of CPC coating which shows a dense and uniform morphology. There are minimal cracks in the coating which shows the coating integrity. The presence of cement hydration products is clearly indicated by

Table 3 Specifications for cement polymer composite coated rebars

Literature/guideline	Coating thickness	Coating continuity
Kamde and Pillai 2020a, b	Thickness to be a minimum of 175 μm	Should be continuous and ineffective surface preparation leads to early corrosion
Yuanliang et al. (2016)	Minimum thickness to be 150 μm	Single layer of 150 μm improved bond strength and thicker layers increased chances of cracking
CSIR- CECRI field trial report	230 ± 66 μm measured thickness	Inadequate application leads to gaps or discontinuity
American concrete institute-ACI 222R	General industrial practise as 250–600 μm	Full coating without any holidays and should avoid any defects

the granular and plate like features. This help in maintain the overall alkalinity of the surrounding coating. The surface preparation plays a significant role in maintaining the coating integrity and adhesion. Sandblasting of the rebar surface prior to the coating application improves the service life and corrosion resistance of the rebar, whereas coating that is applied to the corroded rebar surface may lack the effectiveness and leads to premature corrosion (Kamde and Pillai 2020a, b). Experimental studies using the electrochemical methods also indicated that the CPC coatings applied on the well-prepared surface can delay the corrosion initiation, whereas impact would be reversed in other case (Kamde and Pillai 2022). This points out that effect of CPC coating is not solely based on the composition of coating rather, but also on the quality of application and the surface preparation.

Due to the presence of cementitious materials in the coating, the CPC coating offers a better mechanical performance with the concrete when compared to the pure organic coatings. CPC coated bars without any corrosion exhibited only a slight

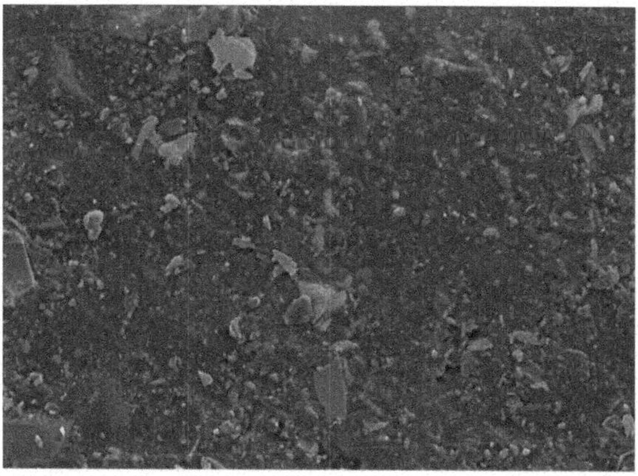

Fig. 9 Microstructural image of CPC coating

reduction in bond strength of approximately 10% due to the relatively smooth surface when compared to the uncoated bars. But, in the case of corroded CPC coated bars, the reduction in the bond strength can go up to 35% (Kamde and Pillai 2018). The cement based epoxy or polymer cement coated rebars are having improved corrosion resistance and longer crack initiation time when compared to uncoated systems which conforms their prolonged service life (Dulaijan et al. 2023). The CPC coatings can promise an effective corrosion resistant coating system in the large scale infrastructures where on-site coating is required if applied effectively.

3.2.4 Emerging Polymer and Nano Coatings

Emerging polymer and nano coatings are the latest advancements in the corrosion resisting coating systems for the reinforcement. These coatings are produced by incorporating nanomaterials, functional polymers and hybrid organic and inorganic materials to enhance, chemical, mechanical and corrosion resistance parameters of the rebars. The conventional coating systems are having some limitations such as the increased susceptibility to mechanical damages, damages induced by the UV exposure, and the problems arising due to the defects in the coatings. The emerging polymer based and nano engineered coatings are having beret barrier resistance, self healing ability and enhanced durability in the exposure conditions. They also reduces the diffusion of chloride ions and moisture through the coating matrix, thus extending the corrosion initiation time and service life of RC structures (Shaker et al. 2022). Polymer coatings are typically prepared from the thermosetting or thermoplastic polymers such as polyurethane, acrylic polymers, epoxy modified resins and silane based coatings. These type of coatings can provide a dense and compact film on the steel rebar and thus improves the adhesion and flexibility. Polyurethane and acrylic polymer coatings provide enhanced resistance to moisture penetration and chloride diffusion inside to the dense polymer network and hydrophobic properties of them (Qian et al. 2025). Silane based polymer coatings can be used rehabilitation and repair works due to their ability to chemically bond and reduce water absorption at the interface.

Nano engineered coatings are developed by mixing nano scale fillers or functional particles into polymer or hybrid matrices inorder to improve the impermeability and better adhesion. The use of nano silica and nano clay fillers improves the compactness and reduces the porosity of the coating matrix due to their enhanced cross linking density. In the graphene based nano coatings, the impermeable graphene sheets creates an effective barrier against the diffusion of the corrosion species. Computational studies also indicates that in the incorporation of graphene into the polymer coatings significantly reduces the diffusion flux of aggressive ions and increases the corrosion initiation time (Shaker et al. 2022). Modified nano coatings with carbon nano tubes are also showing improved mechanical performance and crack resistance, thus maintaining coating integrity and reducing the damage under mechanical loading and environmental exposure. Hybrid coatings are also developed by combining organic polymers with inorganic cementitious or ceramic

Table 4 Thickness and inference of advanced hybrid and nano engineered coatings

Literature	Type of coating	Coating thickness and inference
Suhas and Shanmughapriya (2025)	ZnO-Ni epoxy nanocomposite coating	Thickness of 220 μm and compact morphology post corrosion
Liu et al. (2023)	CeO_2-Graphene oxide/ epoxy nanocomposite coating	Thickness of 100 μm with dense morphology and elevated corrosion resistance
Samad et al. (2023)	SiO_2 and ZrO_2 nanoparticles along with epoxy	Thickness of 100 ± 10 μm, improved hardness and modulus values with corrosion resistance
Fan et al. (2024)	Epoxy combined with graphene nanoplatelets	Thickness of 100 μm, coating resistance decreased and corrosion current density increased with tensile loads

phases. The chemical stability and compatibility of inorganic materials helps in the flexibility and adhesion of the coatings. The electrochemical analysis using accelerated chloride exposure environment confirmed the excellent corrosion resistance of the hybrid organic inorganic coatings. The use of DEIP can result in excellent corrosion inhibiting efficiency in the case of anodic dissolution and as well as cathodic hydrogen evolution (Zoubi et al. 2017). Self healing coatings are another type of engineered coatings that are capable of repairing micro cracks and coating defects by themselves. Accelerated corrosion tests based on ASTM D5894 indicated that mixing self healing microcapsules into the available primers can reduce the underfilm corrosion and preferred method of applying primer would be the spray painting. Such coatings provide an effective solution for long term corrosion protection of reinforcement in aggressive environments. But, they are not capable of repairing the damages that are frequently occurring at the same location (Yimyai et al. 2023). There are need for further long term validation and cost optimization of these coatings before they are implemented in the large infrastructures. Some hybrid and nano engineered coatings along with their thickness and inferences from literature is given in Table 4

4 Mechanism of Protection by Coatings

Corrosion of steel reinforcement within the concrete ius basically an electrochemical process. It involves the anodic oxidation reduction as given in Eq. (1) and cathodic reduction reaction as explained by Eq. (2) previously. When different kind of coatings are applied on the rebar, these electrochemical reactions are further modified based on the protection mechanism that is offered by the coating. The protection mechanism mainly depends on the adhesion and continuity of the coating. The changes in the further electrochemical reactions are governed by the coatings by limiting the ionic transport, discontinuing the electrical connectivity between the anode and cathode or by providing a sacrificial protection to the steel reinforcement.

4.1 Barrier Mechanism

Barrier method of protection mechanism is basically seen in those rebars that are provided with organic coatings such as fusion bonded epoxy coated, cement polymer composite coated and nano polymer coated rebars. They forms a physical barrier, thereby limiting the transport of oxygen, moisture and chloride ions from reaching the steel surface and thereby maintain the alkalinity of concrete. According to the Fick's first law of diffusion (Tutti 1982) as given in Eq. (12) below

$$\mathrm{J} = -\mathrm{D}_{\mathrm{eff}}\frac{\partial C}{\partial x} \tag{12}$$

where, Deff is the diffusion coefficient. When there is the barrier mechanism acting, the organic barrier coatings present reduces and delays the chloride ion penetration to the steel surface. In the case of CPC coatings, the cement polymer film reduces the chloride and moisture ingress, this would result in delaying the corrosion initiation and slows the chloride ingress to the steel surface (Kamde and Pillai 2022). Barrier method of protection also increases the electrical resistivity at the steel surface by lowering the corrosion current. This is evident from the Eq. (13) below.

$$\mathrm{i}_{\mathrm{corr}}\alpha\frac{\beta}{Rp} \tag{13}$$

Here, icorr is the corrosion current density and Rp is the polarization resistance. Both the epoxy coated and the fusion bonded epoxy coating is having high electrical resistivity. This affect the electrochemical reactions on the steel surface. This would inhibit the electrochemical reactions and thereby delaying the corrosion propagation. EIS studies also points that coated rebars are having higher interfacial resistance when compared to that of the uncoated steel reinforcement (Kamde and Pillai 2021). This barrier mechanism thus reduces the transport of chloride ions and improves the service lifer of the structure. But, for the effective protection the coating should be uniform, otherwise it would result in localized pitting corrosion. The cathodic delamination of coated bars is explained in Fig. 10. The cathodic delamination that occurs at the coating may cause the corrosion to progress under the coating.

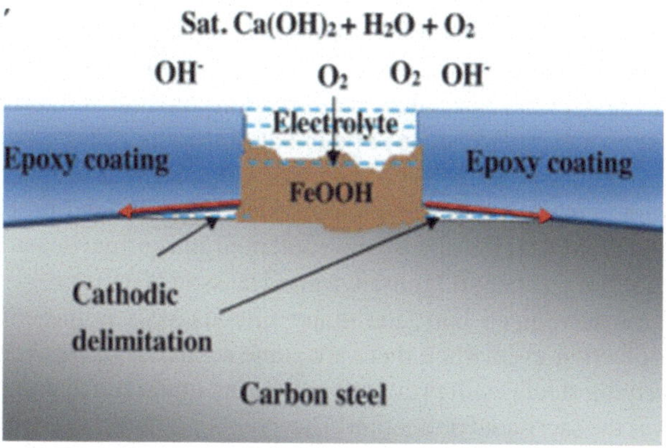

Fig. 10 Cathodic delamination of coated bars (Qian et al. 2025)

4.2 *Sacrificial (Cathodic) Protection*

Sacrificial protection is seen in the steel rebars that are being coated with galvanic protection. The sacrificial protection occurs due to the more electronegativity of the zinc when compared to the uncoated steel rebar. When the zinc coating is exposed to the concrete pore solution, zinc preferentially corrodes, thereby protecting the underlying steel rebar. The electrochemical reactions that are involved includes as given below in Eq. (14) showing anodic dissolution and Eq. (15) showing cathodic dissolution.

$$Zn \rightarrow Zn^{2+} + 2e^- \tag{14}$$

$$O_2 + 2H_2O + 4e^- \rightarrow 4OH^- \tag{15}$$

Electrons that are released during the zinc oxidation acts on the steel surface and prevents the anodic reaction at iron. Thus, steel corrosion is prevented as long as the zinc is available for the sacrificial protection. This series of electrochemical protection is the galvanic protection or cathodic protection and the efficacy of the protection also depends on the type of binder in the concrete (Molina et al. 2017). The zinc further reacts with the hydroxyl ions to form the zinc hydroxide and thus forming calcium hydrozincite and the reaction proceeds is described in Eqs. (15) and (16).

$$Zn^{2+} + 2OH^- \rightarrow Zn(OH)_2 \tag{16}$$

$$Zn(OH)_2 + Ca(OH)_2 \rightarrow Ca[Zn(OH)_3]_2 \tag{17}$$

These corrosion products will fill the pores near the steel–concrete interface and delays the movement of aggressive species. Along with this galvanic coupling also occurs between the steel and zinc, where the corrosion potential of steel reinforcement shifts towards the more electronegative potential of the zinc. This would reduce the thermodynamic tendency for steel oxidation. The potential shift further decreases the corrosion current density and delays the corrosion initiation and propagation. The galvanized bar thus exhibits lower corrosion potential and current densities compared to the uncoated reinforcement (Ogunsanya and Hansson 2018).

In the case of galvanized bars, the major effectiveness is the continuation of the galvanic protection even when there are some scratches or mechanical defects. Here, the underlying steel is still protected as the zinc in the surrounding areas of the defects still gives the sacrificial protection. This points out the robustness of sacrificial protection when compared to other forms of organic coatings and thus reduces the corrosion initiation (Yoon et al. 2024). The zinc corrosion products that are formed also blocks the pores in the concrete and will prevent the movement of ions. It is thus observed in the electrochemical studies that the galvanized bars are having an high resistance when compared to other uncoated bars at the interface due to the corrosion products as well as the reduced electrochemical activity. The presence of zinc coatings delays the corrosion initiation in the chloride exposed environment by providing a sacrificial protection and thus delaying the pitting corrosion. In the carbonation exposed environment, the zinc layer provides the sacrificial protection over the entire length until the zinc coating is significantly consumed. The long term performance of the coating heavily depends on the coating thickness and the exposure conditions. The sacrificial protection and corrosion mechanism of galvanized bar is shown in Fig. 11.

4.3 Inhibitive and Hybrid Mechanism

Hybrid and inhibitive mechanism are in which, the coatings provide both a physical barrier as well as the passive film stabilization of the steel surface. This mechanism typically occurs in cement polymer composite coatings, nano engineered and polymer modified cement coatings. The hybrid coatings interacts with the steel–concrete interface and delays the corrosion initiation by stabilizing the passive film and reducing the movement of ionic species. Hybrid coating systems provide protection through combined effect of reduced permeability, chemical film formation and inhibiting the reactions at the anode. Inhibitive coatings contain chemical species that gets either adsorbed onto the steel surfaces or react with the corrosion products to form a stable protective complex. This can prevent the anodic and cathodic reactions. The adsorbed inhibitor blocks the electrochemically active sites and increases the resistance at the steel concrete interface and thus reduces the corrosion current density (Kamde and Pillai 2022). The coatings also interact with the steel and forms a stable passive layer that further delays the corrosion initiation. Thus in the hybrid and inhibitive mechanism, the coatings protect the steel from oxygen and moisture

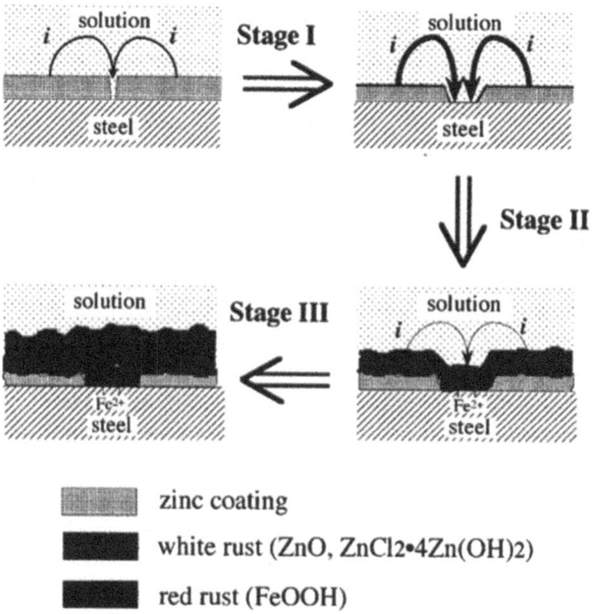

Fig. 11 Sacrificial corrosion mechanism of galvanized bars (Mahdy et al. 2000)

ingress, thereby limiting the anodic and cathodic sites and increasing the electrical resistance during corrosion propagation. But if there are minor defects itself, it will initiate the corrosion faster and thereby reducing the corrosion propagation phase and a significant bond strength reduction can also be observed. Thus in the case of CPC coatings, the performance mainly depends on the coating quality and the chemical reactions at the interface (Kamde and Pillai 2022).

Nano engineered hybrid coatings provides protection by releasing inhibitor species when there is a pH change, coating damage or corrosion initiation. Initially they offers a barrier protection that is followed by a release of inhibitors at the corroded region. The inhibitors then gets adsorbed onto the steel forming protective complex. In the nano particle modified polymer coatings, they contain nanoparticles that fills the pores and reducing the pore connectivity by creating tortuous diffusion paths. This would reduce the porosity and increases the tortuosity of ionic transportation and lowering the diffusion coefficient. In the CeO_2-GO nanocomposite coating, it has improved the coating integrity and reduced ionic permeability resulting in lower corrosion current densities than unmodified epoxy coatings after 60 days of exposure. The EIS results of nano modified epoxy systems also significant increase in the charge transfer resistance (Rct) upto 5–10 times when compared to the other epoxy coatings (Liu et al. 2023). In the self healing coatings, they have microcapsules or nanocontainers that are filled with corrosion inhibitors or healing agents and gets released when there is a microcrack or mechanical damage. The released healing agents repairs the defects in the coatings and provide further barrier protection. The integrated mechanism thus provides superior long term corrosion resistance when compared to the pure barrier or sacrificial protection. Thus, the nano engineered coatings provide protection by lower ionic diffusion, higher polarixzation resistance, by suppressing the anodic reactions and thereby reducing the underfilm corrosion.

4.4 Limitations of Coating Systems

Even though the coated reinforcement systems delay the corrosion initiation and enhances the service life of the structure, their long term performance is affected by issues arising due to coating defects, mechanical damage and electrochemical changes at the point of coating discontinuity. The long term performance of coated rebars are highly influenced by the localized defects even though the overall coating appears to be satisfactory (Qian et al. 2025). The main limitation of the coated rebars is the defects that arise during the manufacturing and handling such as pinholes, holidays and coating discontinuity. Improper surface preparation prior to coating can also lead to weak adhesion and premature corrosion initiation (Kamde and Pillai 2020a, b). Small pinholes or defects can also cause localized corrosion at the region of discontinuity. This happens due to the creation of localized anodic regions where corrosion can initiate and the surrounding coated regions would act as the cathode. In some bars the corrosion starts at the defects and propagates towards the bottom of the coating, generally known as under film corrosion as shown in Fig. 12. This is mainly observed in the organic coatings where the rust products get accumulated under the coating and causes volume expansion, thus exerting pressure on the steel coating interface and causes the removal of the coating. This would result in the exposure of underlying steel and accelerates the corrosion propagation.

Another limitation of the coated bars is the damage that is caused by the bending or handling during the manufacture. The damages caused by bending or handling reduces the corrosion resistance and service life of the coated bars (Kamde and Pillai 2021). The transportation also would cause microcracks to happen on the coated rebars due to the abrasion which acts as the significant corrosion initiation spots. The damaged coated bars thus show higher corrosion current density and corrosion propagation when compared to the uncoated bars pointing the importance of proper handling of the coated rebars. When the coatings are exposed to the alkaline pore solution over a long period, the coating gets damaged due to the chemical degradation and causes the diffusion of oxygen, moisture and chlorides to the steel surface. Carbonation induced alkalinity reduction can also disrupt the passive film over the coating. The electrochemical studies also confirms that gradual reduction in coating resistance over longer exposure duration (Wang et al. 2024). Prolonged exposure of the coated bars to UV radiation causes the initiation of microcracks in the coated rebars. This results in the corrosion initiation and underfilm corrosion at the localized region of microcracks (Kamde and Pillai 2020a, b).

The long term exposure of the galvanized bars results in the consumption of zinc coating. This consumption reduces the coating thickness and causes corrosion of

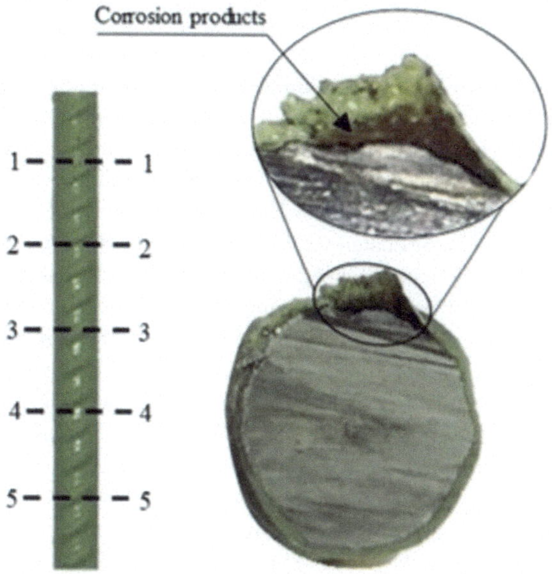

Fig. 12 Underfilm corrosion in FBE coated rebar (Kamde and Pillai 2021)

underlying rebar. The initials reactions between the zinc and alkali pore solution also results in the evolution of hydrogen gas that might affect the interfacial transition zone and bond strength of concrete. The non uniform dissolution of zinc can also localized corrosion and thus care should be taken when using galvanized bars in reinforced structures. Despite all the limitations coated steel rebars delays the corrosion initiation and protects the steel when used without any kind of defects.

5 Performance Evaluation Methods

The corrosion performance of coated rebars is influenced by the electrochemical reactions, diffusion of ions through the coating and the polarization resistance at the steel coating interface. In the evaluation process, coated rebars require specialized evaluation methods due to their barrier techniques, hybrid chemical interaction and sacrificial protection unlike the uncoated bars. The evaluation techniques of coated steel reinforcement include electrochemical methods, accelerated corrosion tests and understanding mechanical performance.

5.1 Electrochemical Techniques

The electrochemical techniques that are commonly used includes the half cell potential (HCP), electrochemical impedance spectroscopy (EIS), linear polarization resistance (LPR). Electrochemical techniques remain as the most effective methods in the performance evaluation and understanding the corrosion performance of steel reinforcements (He 2015). But the conventional electrochemical techniques becomes complex in the coated steel reinforcements as the resistance of coating layer, changes in the ionic path affects the interpretation at the steel concrete interface.

Half cell potential also called as the open circuit potential calculates the corrosion potential of the rebar that is embedded in the concrete. Here, the [potential difference is measured between the reinforcement embedded in concrete and the known reference electrode. The measured potential difference can indicate the corrosion probability of the rebar and is suitable for site as well as laboratory applications (Chandru et al. 2021). The electrochemical reactions are based on the mixed potential theory, where Ecorr is the equilibrium between the anodic and cathodic reactions. As the potential becomes more negative, it indicates the chances of higher corrosion as per the standard in ASTM C876 that relates potential to corrosion probability. But, the HCP data in the coated steel reinforcements may not indicate an ideal data about the corrosion activity. The presence of coating that is providing barrier as protection reduces the ionic conduction and minimizes the potential shift along with the corrosion initiation. In most of the coated rebars, corrosion initiates at the point of defects and the surrounding areas continues the protection, thus resulting in inaccurate potential measurements. Even though, the HCP is effective for the uncoated bars, in the case of uncoated bars the results are not that reliable. The typical half cell potential of uncoated, zinc coated and cement based coating in the in side, transition and outside of slab specimens exposed to slash zone is given in Fig. 13. This indicates that coating does to induce corrosion at the junction of old and repaired structures.

The linear polarization resistance (LPR) is an electrochemical assessment method in which the corrosion rate of steel embedded in the concrete is assessed by applying a potential of about 25 mV around the half cell potential. The LPR is thus obtained by taking the slope along the potential- current density plot around HCP values (Chandru et al. 2021). In the coated reinforcement systems, the polarization resistance has both the coating as well as the charge transfer resistance and has the chances of underestimation of corrosion rate due to the high resistance of the coated rebars. The LPR curves may sometimes fail in indicating the ongoing corrosion when the concrete resistivity is higher or there are more additional layers that are present between coating and steel rebars (Rengaraju et al. 2019). Thus, LPR alone cannot effectively detect the early stage corrosion below the coating layers. The typical LPR curve is shown in Fig. 14. The electrochemical impedance spectroscopy (EIS) provides information about the electrical resistance of the coating. The capacitance indicates the measure of polarization of charges and resistance is the ability to restrict ionic flow (Chandru et al. 2021). EIS gives distinct information about the coating degradation, corrosion kinetics and diffusion processes, thus making it highly effective in

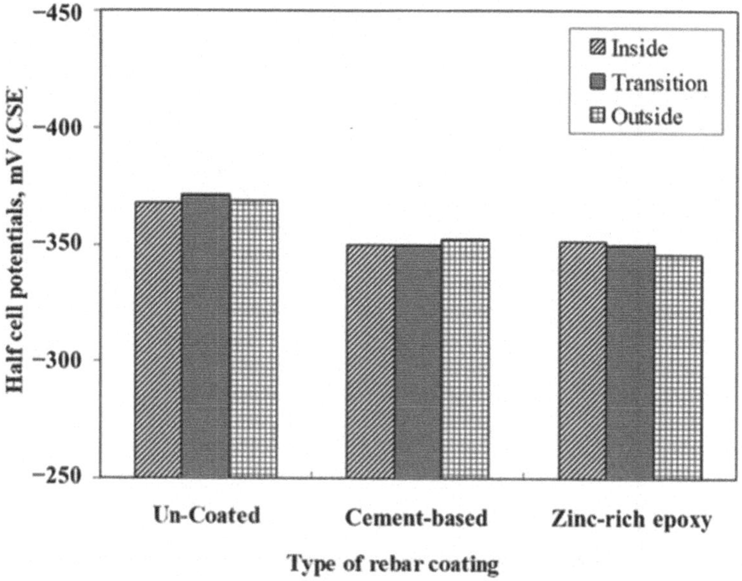

Fig. 13 Half cell potential of coated and uncoated bars (Dulaijan 2023)

evaluation coated reinforcements. The EIS also helps in identifying the early corrosion initiation and as well as the sacrificial or hybrid protection mechanism. EIS can detect corrosion in highly resistive systems due to its advantage to distinguish coating resistance and charge transfer resistance (Rengaraju et al. 2019). The hypothetical EIS response of the coated rebar is shown in Fig. 15 given be. Thus electrochemical techniques acts as the best method of corrosion assessment in coated bars.

5.2 *Accelerated Corrosion Test*

Natural corrosion of the reinforcing steel in concrete is a slow process and in the case of coated rebars it gets increased further. Accelerated testing methods are used to solves this issue and stimulate the long term corrosion of the coated reinforcements in various exposure conditions.

5.2.1 Impressed Voltage

Inn this method of accelerated corrosion testing, an external DC power supply is applied between the anode (coated reinforcing steel) embedded in concrete and the cathode (nichrome mesh) and resulting in the anodic dissolution of iron. The anode and the cathode are partially immersed in the NaCl solution and a DC power supply

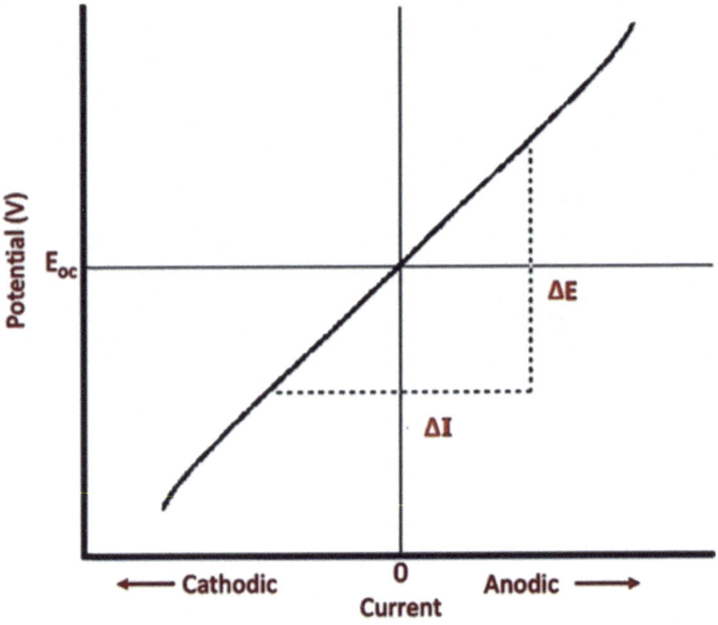

Fig. 14 Typical LPR curve (Daniyal and Akhtar 2020)

Fig. 15 Ideal EIS response of coated steel rebar in cementitious systems (Kamde and Pillai 2021)

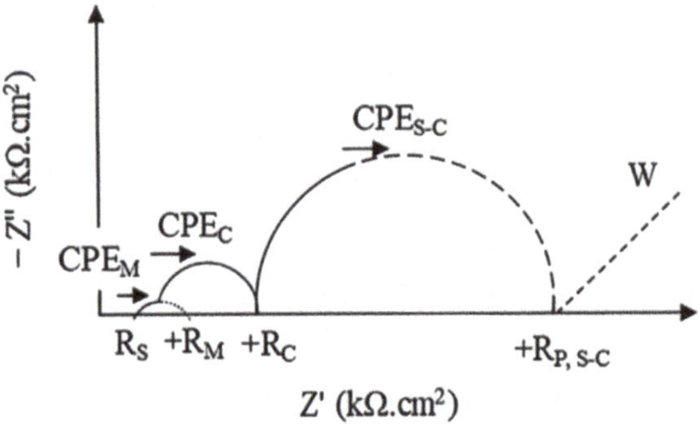

of 5–15 V is applied. The corrosion current is continuously recorded at anode until visible cracks appear on the surface of the specimen (Saraswathy et al. 2017; Chandru et al. 2021). After the cracks are observed on the surface, the rebars are the cleaned of rust using standard procedures and the corrosion rate can be calculated as per the Eq. (18).

$$\text{Corrosion rate } = 87.6 \times \left(\frac{W}{D \times A \times T}\right) \tag{18}$$

By applying the external current, the corrosion initiation is accelerated in the coated rebars which helps in identifying the coating effectiveness. It also serves as a comparative evaluation for different types of coatings under controlled test conditions and also helps in identifying the underfilm corrosion risks. But, here there are chances for the varying current during different stages of corrosion as in the initial stages part of current supplied is used for water electrolysis rather than dissolution of coated rebars (Hong et al. 2020). Also, the corrosion morphology that is produced by forced anodic dissolution can be different from the natural pitting and the electrolysis might change the local Ph, porosity and composition of rust and thus impressed voltage can be best used for comparative studies and not for the absolute field life prediction studies and for this has to be calibrated to field data (Hu et al. 2024).

5.2.2 Chloride Accelerated Exposure Tests

These tests stimulate the natural chloride ingress and thus the corrosion initiation process. In this method, the reinforced specimens are immersed in NaCl solution and then exposing to accelerating wet and dry cycles. During the cyclic exposure chloride ions penetrate inside the concrete and initiates the corrosion. The corrosion initiation and propagation can be assessed by electrochemical techniques such as OCP, EIS and LPR. The corrosion initiation occurs according to the Ficks's second law of diffusion as given in Eq. (12). By further increasing the chloride concentration and moisture availability, the rate of diffusion and chloride flux can be increased and the rate of transport-controlled corrosion initiation and propagation of the coated bars can be identified. The major limitation is that the test is sensitive to cycle details and the drying of specimen and has less control on the corrosion current during the propagation stage (Ali et al. 2024).The accelerated corrosion setup using three electrode system is shown in Fig. 16.

5.2.3 Macrocell Test

This accelerated corrosion test is done as described in ASTM G109-13. It determines the chemical induced corrosion resistance of the reinforcement inside the concrete. The specimens are prepared according to the specifications as shown in Fig. 17. The test is started after 28 days of curing The ratio of anode to cathode is taken as 1:2. Salt ponding is done on the G109 samples and they are subjected to alternate wetting and dry cycles. In the wet cycle (15 days), 3% NaCl solution is poured in these ponds, followed by the dry cycle. At the end of every wet cycle marcocell current (μA) is measured using a voltmeter. In the case of coated bars there are chances for the marcocell to occur within the same bar that are having some defects or holidays (Kamde et al. 2021). As per ASTM G109 standard, when the total corrosion was equal to or greater than 150 C, the rebars are said to be corroded.

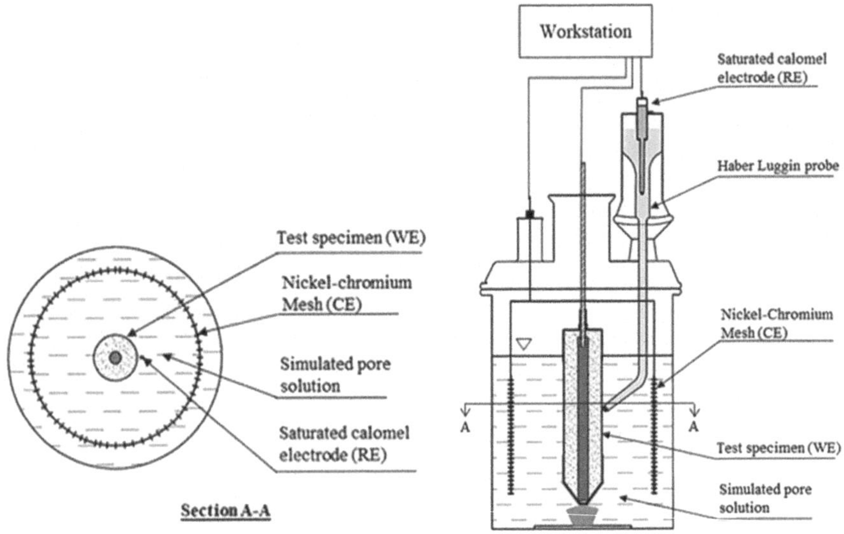

Fig. 16 Accelerated corrosion test using three electrode setup (Rengaraju et al. 2019)

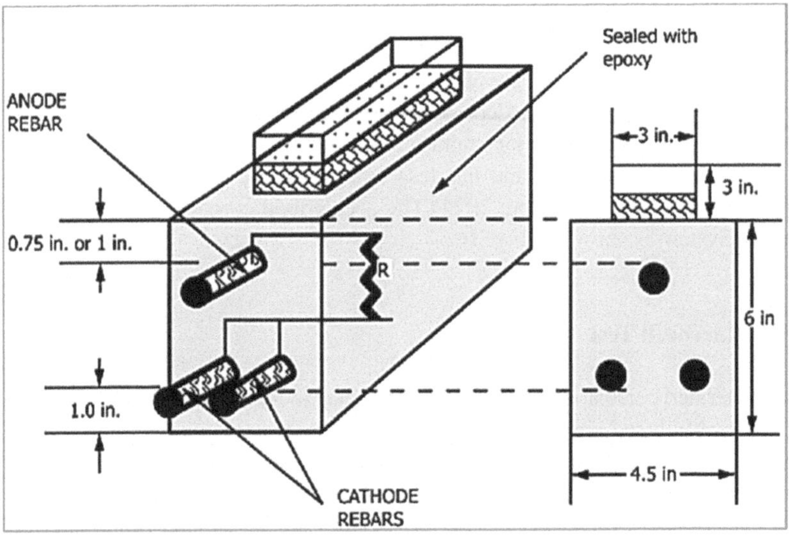

Fig. 17 Specimen specifications for macrocell corrosion test (ASTM G109-13)

Macrocell testing evaluates the chloride threshold of coated steel reinforcement under accelerated chloride exposure and microcell conditions. Here, the electrochemical coupling between the anode and cathode is simulated as the current flow is measured thereby indicating the corrosion initiation. Macrocell current can range upto several hundreds of μA depening on chloride gradient and resistivity of the concrete as higher current indicates the faster anodic dissolution at the contaminated zone. This test helps in assessing the impact of the localized corrosion also in the coated bars along with the corrosion initiation process. There are chances that marcocell distribution can increase the damage the anodic patches and increses the rate of local loss more than the uniform loss (Jian et al. 2021). ASTM G109 helps in long term monitoring of corrosion behavior of coated steel bars under wet and dry exposure cycles and quantifies the effect of current drawn by the local anode that is formed at location of holiday in the coating.

5.3 Bond Strength and Mechanical Performance

The coatings are provided to the steel reinforcement inorder to increase the corrosion free service life. But the presence of coating can influence the mechanical properties at the steel concrete interface, thus affecting bond strength, slip and the cracking response. The presence of coatings can influence the surface roughness and the adhesion between steel and concrete, and thus influences the bond stress- slip relations of the reinforced concrete (Qian et al. 2025) The steel concrete bonding is influenced by the chemical adhesion, frictional resistance and mechanical interlocking of the reinforcement ribs with the concrete and there are chances for the coatings applied to make changes in this. Organic coatings generally reduces the mechanical interlocking and friction due to their smooth polymer surfaces, which eventually leads to decreased bond strength when compared to the uncoated bars. But the cement particles in the cement polymer composite coating may provide chemical adhesion with the concrete. CPC coated bars exhibited only a slight reduction in bond strength when compared to the uncoated bars pointing the structural integrity of the coated bars. But when there is even a slight corrosion, the bond strength has reduced drastically to about 30% indicating the effect of corrosion on the bond performance (Kamde et al. 2018).

The epoxy coated bars are having slightly lower bond strength when compared to the uncoated bars due to their smooth surface, but their improved corrosion resistance overperforms this bond strength reduction. In the case of metallic coated bars such as galvanized bars, they generally exhibits comparable bond strength due to the formation of calcium hydrozincite at the steel concrete interface, even though early age hydration reduces the bond strength slightly due to pore formation (Pokorný et al. 2022). Corrosion also significantly alters the bond strength. Corrosion results in the formation of expansive rust formation that induces radial tensile stresses in concrete cover, thereby leading to cracking and loss of bond strength. The corrosion initiation and propagation mechanism would strongly influence the bond behaviour and mechanical performance of the coated rebars (Kamde et al. 2021). Coatings applied on the steel reinforcement influences the mechanical properties such as yield strength, ductility and structural performance. Most of the coatings are very thin and do not significantly alter the mechanical properties of the coated rebars. But thicker coatings and the defects in the coatings can influence the stress redistribution and causes localized cracking to occur. Modern coating systems such as waterborne epoxy, moisture cure polyurethane and UHPC coatings shows improved bond preservation compared to the traditional protection with bond stress reduction ranging from 9 to 16% (Vanamana et al. 2024). There should be critical evaluation of the mechanical performance of the coated rebars along with corrosion assessment inorder to ensure long term durability of the structure.

6 Durability Performance in Aggressive Environment

The long-term durability of the coated reinforcement is influenced by various exposure conditions such as the rate of carbonation, chloride concentration, humidity and variation sin the temperature. The exposure of reinforcements to these aggressive environments can accelerate the rate of corrosion. The various coating systems applied to the reinforcement provides protection in these aggressive environments by passive film stabilization, reducing ionic transport and further delaying the initiation of corrosion.

6.1 Marine and Coastal Conditions

The reinforced concrete structures exposed to the marine and coastal conditions are under severe exposure conditions due to the continuous chloride penetration from the seawater, effect of alternate wet and dry cycles during the tides and the chlorides and other chemicals that are present in the air. Due to faster diffusion of the chloride ions through the concrete cover, tye corrosion initiates at a faster rate in those structures that are exposed to sea water. This method of diffusion is explained by the Fick's second law of diffusions and the presence of coatings can reduce this to a greater extent. The presence of coatings can indirectly reduce the effective chloride ingress by controlling the moisture ingress and and increasing the cvhloride threshold at the steel- coating interface. When coating systems are provided in low permeable concrete, diffusion coefficient have reported to be less than 50–80% (Wang et al. 2024). Fusion bonded epoxy coated rebars delays the corrosion initiation when compared to the uncoated rebars during the field exposure studies. But when there are chances of any defects or discontinuities in the coating, localized corrosion can initiate at these locations (Qian et al. 2025). Similarly, when there is proper surface preparation in the cement polymer coated rebars, they also delays the corrosion initiation under aggressive environment conditions (Kamde and Pillai 2021).

Reinforced concrete structures that have galvanized reinforcement also exhibits improved durability in coastal exposure due to the sacrificial protection of the zinc that protects the underlying reinforcement, even though there are some minor cracks on the coatings (Pokorný et al. 2024). Once the depassivation has started the corrosion current density for galvanized coating (< 1 μA/cm^2), is less than epoxy coated (1–5 μA/cm^2), nano modified coatings (< 2μA/cm^2) and uncoated bars (5–20 μA/ cm^2) due to the sacrificial protection of zinc coating (Wang et al. 2024). For the long term performance, it would be suggested to use the galvanized bars without any defects. In overall, during the marine exposure conditions, coated rebars significantly improve the durability by delaying chloride induced corrosion initiation and propagation rates, but the performance also depends on the coating uniformity, adhesion and the wet dry cyclic exposure cycles.

6.2 *Industrial and Urban Exposure*

In industrial and urban environments, reinforced concrete structures are exposed to higher levels of carbon, nitrogen and sulphur oxides along with changes in humidity and carbonation is the primary cause of corrosion in such environments. CO_2 reacts with the calcium hydroxide in the concrete, reduces the alkalinity and makes the passive film unstable and resulting in the in the initiation of corrosion. The schematic representation of carbonation of concrete is given below in Fig. 18. Coatings when provided to the reinforcement lower corrosion current densities and delay corrosion initiation and propogation when compared to the normal uncoated reinforcements (Shetty et al. 2022). The typical corrosion current densities of uncoated rebar (1–5 μA/cm^2) is higher than that of epoxy coated (< 1 μA/cm^2) and nanomodifed (< 0.5 μA/cm^2). But long term carbonation can further lead to progressive steel loss.

The barrier type of coatings provided to the reinforcement limits the access of further moisture and oxygen to the steel reinforcement surface and can slow down the corrosion propagation even the carbonation front has started causing rebar depassivation. The sacrificial protection of galvanized reinforcement also provides the

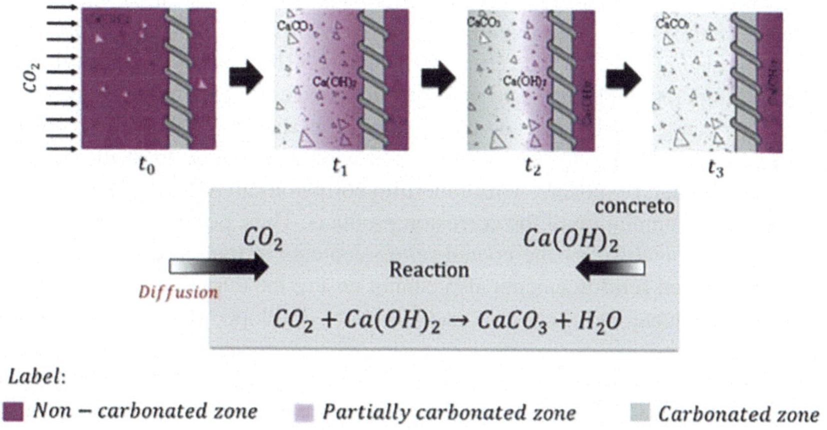

Fig. 18 Schematic representation of carbonation of concrete (Felix et al. 2017)

necessary protection in the carbonation environment thus delaying the corrosion initiation and preventing depassivation of the reinforcement. Hybrid coatings maintain a more stable interfacial environment and improves the passive film stability through chemical interaction with the cementitious materials (Kamde and Pillai 2021). Thus, the coated reinforcement systems improve the durability and help in preventing carbonation induced depassivation, thereby reducing corrosion initiation and propagation.

6.3 Long Term Field Performance

Long term field exposure study serves as the most reliable method of durability performance assessment of coated steel reinforcement under the real time environmental conditions. When the bridge decks and marine infrastructures has been monitored over a long term, it indicates that the coated reinforcement significantly delays the corrosion initiation and thereby enhances the service life of the structure. But there were some localized corrosion that was observed due to the coating defects in the reinforcement (Qian et al. 2025). The long term service life is also higher for epoxy coated (50–70 years), galvanized (60–90 years), CPC or nano modified (60–80 years) when compared to the uncoated (25–40 years) under the same marine exposure conditions (Qi et al. 2026; Wang et al. 2024).

Galvanized reinforcement also has long tern performance due to the sacrificial protection of the zinc coating. This continues as long as the zinc coating is available and the long term performance also depends on the zinc layer thickness. This is due to the accelerated consumption of zinc in the highly aggressive environments significantly affects the coating protection. CPC coatings can delay the corrosion initiation in chloride contaminated concrete and improve the service life of the structure when proper surface preparations and coating application is followed on the reinforcement (Kamde and Pillai 2021). This puts an emphasis on the importance of quality control during the manufacture process. Environmental degradation of the coating also affects the long term performance. This mainly includes the UV exposure, temperature changes and mechanical damages that are induced during the service time can gradually degrade the coating properties and affects the corrosion performance over the time. The localized corrosion can also occur at the coatings during the prolonged exposure. This underfilm corrosion can remain undetected and results in the accumulation of the corrosion products. Thus, periodic inspection and electrochemical analysis of the coated reinforcements in the structures should be conducted. Coated reinforcements also cannot ensure the durability if the concrete quality is inadequate and there are cracks that are developed due to the structural loading issues.

The long term field performance points out that coated reinforcement provide significant durability of structure by delaying corrosion initiation and propagation and thereby extending the service life of the structure. But their effectiveness solely depends on the coating uniformity, environmental exposure conditions and long term degradation mechanism along with the surface preparation. For the extended service life, the performance of the coated rebars are to be monitored and effectively maintained. Thus, the coated bars enhances the service life by reducing effective chloride diffusion, increasing the chloride threshold levels, lowering corrosion current density and extending corrosion initiation time by 2–3 times, thereby increasing the total service life by at least 20–50 years.

7 Service Life and Sustainability Considerations

7.1 Service Life Estimation

Service life is the period during which the structure performs effectively without any defects and doses not involve any maintenance and repair costs. The service life is divided into two such as corrosion initiation and propagation phase. The initiation period of corrosion is modelled using the Fick's second law of diffusion. The durability design based approaches mainly focuses on the delaying the corrosion initiation phases. Service life models such as Life-365 and fib Bulletin-34 uses the diffusion parameters, cover depth, temperature and exposure conditions inordeer to predict the corrosion initiation time and predict the service life of the structure. The presence of coatings influences these service life prediction models by lowering the chloride diffusion coefficient and increasing the time for corrosion initiation through barrier resistance, sacrificial or hybrid protection mechanisms. The total service life can be defined as shown below.

$$t_{service} = t_{initiation} + t_{prpogation}$$

In addition to this, the coatings may also influence the propagation phase by reducing the corrosion current density and delaying the anodic dissolution of the reinforcement, thus delaying the cracking and structural deterioration. Service life prediction models generally consider the coating resistance as an additional transport resistance layer. In addition to transport resistance, coatings also influences the service life by altering the electrochemical reactions at the steel concrete interface. Service life prediction also points that considering the crack propagation can also increase the predicted service life. Barrier protection methods can enhance the service life by delaying the corrosion in chloride contaminated environments when the coating application and surface preparation are properly done (Kamde and Pillai 2021). In metallic coatings such as galvanized bars, sacrificial protection is offered, and this maintains the steel in a cathodic state even though local damage has occurred. Coating defects such as holidays or microcracks may reduce the effective service life when compared to those in the case of ideal coatings. When considering the durability based service models, they also include the resistivity, permeability and degradation of the coating into consideration. Such models points out that there are chances for the coating properties to evolve over the time and may reduce the efficiency of protection and accelerates the chances of corrosion. The service life estimation of coated bars should and consider both initial coating performance and its time dependent degradation. Thus considering the whole scenario, coated reinforcements can significantly enhance the service life and there should be a probabilistic modelling including the coating parameters for the service life estimation.

7.2 *Cost Implications*

Corrosion of reinforcement in structures leads to a huge economic loss worldwide including the repair, maintenance and rebuilding costs. Although, the coated reinforcements used in reinforced concrete has a high initial material and application cost, but when considering the long term effect they are economically beneficial. This is mainly due to the reduced repair, maintenance and repair costs when compared to those of uncoated. From the Life cycle cost analysis it is evident that the corrosion protection measures as in the case of coated rebars significantly reduces the life cycle expenditure by minimizing the repair frequency and extending the service life of structure. As the coated reinforcement extend the corrosion initiation time, they reduce the number of repair cycles and thus results in the life cycle cost (Qian et al. 2025). Fewer repair costs not only reduces the direct repair cost, but also the indirect costs such as those associated with reconstruction and operational downtime of structure. Different coating systems are having different cost considerations. In the case of organic coatings, they are relatively economical and hence widely used in the reinforced concrete structures. Metallic coating such as the galvanized, due to the zinc content has a higher initial cost, but in the long run due to the additional sacrificial as well as barrier potential, the initial cost is compromised. Though the hybrid coatings have an additional applying and handling costs, but they also provide the enhanced durability due to the interfacial compatibility and reduces the life cycle cost.

The cost effectiveness of coatings also depends on the exposure conditions. In those environments with low corrosion risk additional coating cost may not have economic benefits, but in the aggressive environment such as marine areas, coated reinforcement are advantageous. Thus the cost implication of the coated reinforcement should be valuated from a life cycle perspective rather than the material costs. Therefore adopting reinforcements with coating is an economical durability strategy that would enhance the longevity of the structure and overall life cycle cost of the reinforced strcture.

7.3 *Sustainability Aspects*

Durability is one of the main strategies for reducing the environmental footprint of concrete structures. Using coated reinforcement increases the service life of the structure. This would reduce the environmental impact that is associated with repair, reconstruction and additional consumption of various energy sources as the steel manufacturing releases a large amount of CO_2 to the environment. By delaying the corrosion initiation and propagation, the coated reinforcement indirectly reduces the carbon emission and thus contributes to the reduction in carbon footprint of concrete and limiting the resource consumption for developing new structures.

Using coated reinforcement along with low-permeability concrete, adequate cover thickness, and proper curing practices provides a durable design that enhances both service life and sustainability. The life cycle assessment based on durability indicates that extending service life significantly reduces overall environmental impacts. The sustainability aspects of coated reinforcement are thus basically dependent on their ability to enhance the service life, reduce maintenance, and minimize the environmental impacts that are associated with corrosion.

8 Conclusions

From the above discussion on coated steel reinforcement following conclusions can be made:

- Corrosion initiation in reinforced structures is basically governed by chloride diffusion and the critical chloride threshold. Uncoated reinforcement may depassivate within 15–25 years with a diffusion coefficient of Deff = 1–2 × 10–12 m2/s and reducing the Deff can significantly reduce the corrosion initiation. This highlights the significance of corrosion resistance strategies that limits the transport of moisture and chloride.
- Coated reinforcement can significantly increase the corrosion initiation time by 2–4 times and increase the service life of the structure upto 3–6 times. Such elay in the corrosion initiation can reduces the repair of structures and helps in shifting the durability design from repair focused to design life improvable performance.
- The performance of coated rebars can be assessed through various electrochemical techniques. These electrochemical measurements (OCP, EIS,LPR) indicates the corrosion current density of coated bars upto 60–95% when compared to uncoated rebars. While the uncoated steel have an icorr value of about 2–5 $\mu A/cm^2$, for the coated bars it remains as low as 0.2–0.5 $\mu A/cm^2$ during the same period under chloride exposure and reduces the propagation rate and development of crack.

- Bond strength of the coated bars is comparable to that of uncoated steel rebars when there are no defects in coatings, but significantly reduces with corrosion initiation or defects in concrete, and when there are changes in interfacial friction and mechanical interlock. Galvanized bars have zinc corrosion products that enhance the mechanical interlocking, along with providing sacrificial protection.
- Though the coated reinforcements have a higher initial material cost of about 15–40% than uncoated bars, the life cycle cost of the coated over a period of about 60–80 years shows a total cost reduction of 15–30% when exposed to the aggressive environments. The lack of major repair and rehabilitation works within the first cycle of about 20–25 years itself significantly reduces the life cycle cost. Break-even time can typically be achieved within 12–20 years for coated bars without any defects exposed to aggressive environments.
- Steel production emits approximately 1.8–2.2 t CO^2 per tonne of steel. The uncoated bars have a shorter service life and require either maintenance or replacement that effectively doubles the carbon impact. Coated reinforcements that extend the service life when compared to uncoated can reduce the cumulative CO^2 emissions by 25–45%, mainly by avoiding the replacement of rebar or material during repair and associated construction costs.
- Probabilistic durability models can further indicate that the reliability index (β) for the uncoated reinforcement in aggressive exposure can drop below 1.5 within 25 years of service life. Whereas, in the coated systems reliability index (β) remains $\geq$ 3.0 for over 50 years when bars are perfect without any holidays or defects. This indicates that the coatings not only delay the corrosion initiation but also robust the service life performance.

Acknowledgements The authors gratefully acknowledge the financial support provided by the Anusandhan National Research Foundation (ANRF), Government of India, under Project Sanction No. ANRF/IRG/2024/000739/ENS, titled "Modelling the Electrochemical Responses and Understanding the Corrosion Mechanisms of Coated Steel Bars Embedded in Reinforced Concrete Systems."

References

Al-Dulaijan SU (2023) A laboratory and field assessment of the performance of rebar coatings. Materials 16(12):4270

Ali M et al (2024) A review on chloride induced corrosion in reinforced concrete structures: Lab and in situ investigation. RSC Adv 14(50):37252–37271

Angst U, Büchler M (2020) A new perspective on measuring the corrosion rate of localized corrosion. Mater Corros 71(5):808–823

Brindha U, Maheswaran J, Chellapandian M, Arunachelam N (2023) Quantitative assessment of strengthening strategies and design recommendations for the repair of corrosion-damaged reinforced concrete members. Buildings 13(4):1080

Chandru P, Karthikeyan J, Natarajan C (2021) Techniques to assess the corrosion resistance and corrosion rate of the steel embedded in concrete. Building Pathologies and Acoustic Performance. Springer International Publishing, Cham, pp 33–54
Daniyal Md, Sabih A (2020) Corrosion assessment and control techniques for reinforced concrete structures. New challenges and industrial applications for corrosion prevention and control. IGI Global Scientific Publishing, pp 226–259
Das CS, Zheng H, Dai JG (2025) A review of chloride-induced steel corrosion in coastal reinforced concrete structures: Influence of micro-climate. Ocean Eng 325:120794
Dhanya BS, Rathnarajan S, Santhanam M, Pillai RG, Gettu R (2019) Carbonation and its effect on microstructure of concrete with fly ash and ground granulated blast furnace slag. Indian Concr J 93(4):10–21
El-Mahdy GA, Nishikata A, Tsuru T (2000) Electrochemical corrosion monitoring of galvanized steel under cyclic wet–dry conditions. Corros Sci 42(1):183–194
Fan L, Hou G, Liao X, Shang H, Liu N, Lu D, Sun C et al (2024) Corrosion behavior of modified cement-based coated steel rebar subjected to different levels of tensile loads. J Build Eng 91:109707
Felix EF, Carrazedo R, Possan E (2017) Parametric analysis of carbonation process in reinforced concrete structures through artificial neural networks. Rev ALCONPAT 7(3):302–316
Harilal M, Anandkumar B, Lahiri BB, George RP, Philip J, Albert SK (2020) Enhanced biodeterioration and biofouling resistance of nanoparticles and inhibitor admixed fly ash-based concrete in marine environments. Int Biodeterior Biodegradation 155:105088
Harilal M, Kamde DK, Uthaman S, George RP, Pillai RG, Philip J, Albert SK (2021) The chloride-induced corrosion of a fly ash concrete with nanoparticles and corrosion inhibitor. Constr Build Mater 274:122097
He Q, Zhang D, Huang Y, Yang Y, Ma G (2025) Recent progress of corrosion prevention method of magnesium alloy. Steel Res Int 96(2):2400376
He B, Han PJ, Lu C, Bai XH (2015) Effect of particle size on the electrochemical corrosion behavior of X70 pipeline steel: Einfluss der Partikelgröße auf das elektrochemische Verhalten von X70 Pipelinestahl. Materialwissenschaft und Werkstofftechnik 46(11):1077–1087
Hong S, Zheng F, Shi G, Li J, Luo X, Xing F, Dong B et al (2020) Determination of impressed current efficiency during accelerated corrosion of reinforcement. Cement Concr Compos 108:103536
Hu J, Liu D, Deng P, Shangguan J, Zheng G, Yang J (2024) Research on the optimal protection parameters of graphene composite conductive coatings combined with impressed current cathodic protection technology in marine atmospheric environments. Coatings 14(10):1263
Ibrahim M, Adewumi AA (2017) Studies on the performance of migratory corrosion inhibitors in the corrosion control of concrete rebars. University of Hafr Al-Batin, Saudi Arabia
Jiang Z, Li S, Fu C, Dong Z, Zhang X, Jin N, Xia T (2021) Macrocell corrosion of steel in concrete under carbonation, internal chloride admixing and accelerated chloride penetration conditions. Materials 14(24):7691
Joseline D, Kamde DK, Rengaraju S, Pillai RG (2019) Residual service life estimation and its importance for pretensioned concrete (PTC) bridges in coastal cities
Kamde DK, Pillai RG (2017) Comparison of corrosion of damaged fusion-bonded-epoxy-coated (FBEC) and uncoated steel rebars. 71st RILEM Annual Week. RILEM, Chennai, India
Kamde DK, Pillai RG (2018) Effect of the degree of corrosion on bond performance of cement polymer composite (CPC) coated steel rebars. In: MATEC web of conferences, vol 199, p 04010. EDP Sciences
Kamde DK, Pillai RG (2020a) Effect of surface preparation on corrosion of steel rebars coated with cement-polymer-composites (CPC) and embedded in concrete. Constr Build Mater 237:117616
Kamde DK, Pillai RG (2020b) Effect of sunlight/ultraviolet exposure on the corrosion of fusion-bonded epoxy (FBE) coated steel rebars in concrete. Corrosion 76(9):843–860
Kamde DK, Pillai RG (2021) Corrosion initiation mechanisms and service life estimation of concrete systems with fusion-bonded-epoxy (FBE) coated steel exposed to chlorides. Constr Build Mater 277:122314

Kamde DK, Kessler S, Pillai RG (2021) Condition assessment of reinforced concrete systems with fusion bonded epoxy coated rebars. Corrosion 77(12):1332–1343

Kamde DK, Pillai RG (2022) Corrosion initiation and its effect on bond characteristics and service life of reinforced concrete systems with Cement-polymer-composite coated steel rebars. In: Structures, vol 44. Elsevier, pp 248–260

Liu X, Wu Z, Lyu Y, Li T, Yang H, Liu Y, Shah SP et al (2023) Corrosion resistance of CeO_2-GO/Epoxy nanocomposite coating in simulated seawater and concrete pore solutions. Polymers 15(12):2602

Mehta PK, Monteiro PJM (2006) Concrete-microstructure, properties, and materials, 3rd edn. The McGraw-Hill Companies Inc., New York

Ming J, Wu M, Shi J (2021) Passive film modification by concrete carbonation: re-visiting a corrosion-resistant steel with Cr and Mo. Cement Concr Compos 123:104178

Molina FL, Alonso MA, Moreno MS, Centenero RJ (2017) Corrosion protection of galvanized rebars in ternary binder concrete exposed to chloride penetration. Constr Build Mater 156:468–475

Mukhopadhyay A, Sahoo S Surface engineering of construction steel for corrosion protection–a review

Mukhopadhyay AK, Mistri A (2025) Evaluate performance of sealers and coatings applied to txdot bridge substructures

Nong Y, Chen Z, Yi C, Wang Y (2025) New insight into the chloride-induced depassivation of steel bar in concrete pore solution: The retarding role of calcium component in passive film. Constr Build Mater 502:144338

Ogunsanya IG, Hansson CM (2018) The influence of coating thickness and composition on the corrosion propagation rates of galvanized rebar in cracked concrete. Corrosion 74(1):134–143

Osibodu SJ, Akerele DD, Bamigboye GO (2025) Efficacy of self-healing concrete for mitigating reinforcement corrosion: a critical review of transport properties and electrochemical performance. Buildings 15(21):3875

Patnaik A, Stroia M M (2019) Structural and corrosion performance of continuous galvanized rebar (CGR). GalvaBar: Catoosa, OK, USA

Pokorný P, Kostelecka M, Prodanovic N, Sýkora M (2022) Effect of calcium hydroxyzincate on bond strength of hot-dip galvanized plain bars with normal strength concrete. Cement Concr Compos 130:104540

Pokorný P, Chobotský T, Prodanovic N, Steinerová V, Hurtig K (2024) Bond strength and corrosion protection properties of hot-dip galvanized prestressing reinforcement in normal-strength concrete. J Compos Sci 8(10):407

Qi H, Chen X, Fu S, Wu Y, Hu F, Hui L, Shi J (2026) Numerical simulation of pipeline corrosion via coupled nonisothermal turbulent heat transfer and electrochemical models. Mater Corros 77(3):372–390

Qian H, Mao J, Ye S, Xu Z, Chen S, Liu Y, Yan D (2025) Coated rebars for reinforced concrete structures: a review. J Build Eng 109:113000

Rathnarajan S, Pillai RG (2025) Carbonation models using mix-design parameters for concretes with supplementary cementitious materials. J Build Eng 104:112392

Rengaraju S, Neelakantan L, Pillai RG (2019) Investigation on the polarization resistance of steel embedded in highly resistive cementitious systems–An attempt and challenges. Electrochim Acta 308:131–141

Rengaraju S, Pillai RG (2021) An accelerated chloride threshold test for uncoated steel in highly resistive cementitious systems (hr-ACT test). Constr Build Mater 305:124797

Rengaraju S, Pillai RG, Gettu R (2023) Input parameters and nomograms for service life-based design of reinforced concrete structures exposed to chlorides. In; Structures, vol 56. Elsevier, p 104847

Samad UA, Alam MA, Abdo HS, Anis A, Al-Zahrani SM (2023) Synergistic effect of nanoparticles: enhanced mechanical and corrosion protection properties of epoxy coatings incorporated with SiO2 and ZrO2. Polymers 15(14):3100

Sánchez Montero J, Saura Gómez P, Torres Martín JE, Chinchón-Payá S, Rebolledo Ramos N (2023) Variation of corrosion rate, vcorr, during the carbonation-induced corrosion propagation period in reinforced concrete elements. Materials 17(1):101

Saraogi A, Warudkar V (2018) Study of performance parameters of fusion bonded epoxy coated reinforcing bars and their testing. Mater Today: Proc 5(2):4138–4148

Saraswathy V, Karthick S, Lee HS, Kwon SJ, Yang HM (2017) Comparative study of strength and corrosion resistant properties of plain and blended cement concrete types. Adv Mater Sci Eng 2017(1):9454982

Shokri A (2025) Reinforced concrete corrosion; Mechanism, challenges, prospective, and future roadmap. J Electrochem Sci Eng 15(5)

Shaker M, Salahinejad E, Cao W, Meng X, Asl VZ, Ge Q (2022) The effect of graphene orientation on permeability and corrosion initiation under composite coatings. Constr Build Mater 319:126080

Shetty S, Banerjee S, Tallur S, Desai YM (2022) Effectiveness of multi-physics numerical model in simulating accelerated corrosion with spatial and temporal non-uniformity. Surf Topogr Metrol Prop 10(4):045012

Suhas VS, Shanmugapriya T (2025) Synergistic Zno-Ni/Pvdf-Epoxy nanocomposite coatings for enhanced corrosion resistance and mechanical performance of steel reinforcements in concrete structures. Results Eng 107296

Tutti K (1982) Corrosion of steel in concrete. Swedish Cement and Concrete Research Institute, of CBI research, pp 82–84, 468

Vanama RK, Ramakrishnan B (2024) Experimental investigation on the bond-slip behavior of rebars coated with epoxy and polyurethane systems: a comparative study. J Build Eng 94:110066

Wang Q, Wang Z, Li C, Qiao X, Guan H, Zhou Z, Song D (2024) Research progress in corrosion behavior and anti-corrosion methods of steel rebar in concrete. Metals 14(8):862

Yimyai T, Crespy D, Rohwerder M (2023) Corrosion-responsive self-healing coatings. Adv Mater 35(47):2300101

Yoon IS, Nam JH (2024) Evaluation of corrosion resistance performance of anti-corrosion reinforcements with surface damage. KSCE J Civ Eng 28(11):5189–5202

Yuanliang X, Kunrong W, Zhiyong L, Zhengguang Y (2016) Effect of coating thickness on bond behaviors of polymer cement coated plain steel bar with concrete and finite element modeling. The Open Civ Eng J 10(1)

Zoubi WA, Min JH, Ko YG (2017) Hybrid organic-inorganic coatings via electron transfer behaviour. Sci Rep 7(1):7063

Corrosion Performance and Long-Term Durability of Reinforced Limestone Calcined Clay Cement (LC3) Concrete

R. Anagha and Chandru Pichaimuthu

Abstract A new method with less clinker, limestone calcined clay cement (LC3) concrete is becoming popular because of its excellent mechanical performance, low carbon emissions and sustainable material composition. LC3 lowers the embodied carbon of concrete while maintaining good strength and workability compared to conventional Portland cement systems. In addition to its environmental benefits, LC3 has promising short-term durability performance, particularly in chloride-rich environments. The mechanism of corrosion in steel embedded in LC3 concrete is covered in this chapter, along with how the steel behaves in chloride and carbonation environments. Additionally covered are the modifications to the microstructure and transport mechanism. Although the microstructure of LC3 systems is thought to be more refined, the lower portlandite content in its pore solution is not advantageous from the perspective of carbonation. The acceptance of LC3 in structural applications depends on its long-term durability as well as its field performance.

Keywords Limestone calcined clay cement · Microstructure · Chloride attack · Carbonation · Electrochemical assessment

1 Introduction

One of the main industries contributing to carbon emissions is the cement manufacturing sector. SCMs, or supplementary cementitious materials, have a significant potential to lower emissions and carbon footprint. Fly ash, slag, metakaolin, rice husk ash, and other major SCMs are among them, but because their supply is diminishing, alternative sources are being looked into (Scrivener et al. 2018) (Ez-zaki et al. 2021). Limestone calcined clay cement systems are gaining popularity in recent times,

R. Anagha · C. Pichaimuthu (✉)
Department of Civil Engineering, National Institute of Technology Calicut, Calicut, Kerala, India
e-mail: chandru@nitc.ac.in

R. Anagha
e-mail: anagha_p240509ce@nitc.ac.in

João M. P. Q. Delgado (ed.), *Advanced Materials, Corrosion Science, and Rehabilitation of Concrete Infrastructure*, Building Pathology and Rehabilitation 29,
https://doi.org/10.1007/978-3-032-28851-6_2

mostly because of their low carbon emissions, good short-term performance in laboratory experiments, and the availability of raw materials. Limestone calcined clay cements are made by substituting a mixture of limestone and calcined clay with roughly 5% gypsum for 40–50% of the clinker content. The process significantly lowers carbon emissions because it involves calcining clays at a temperature of 700–850 °C, which is nearly half of the clinkering temperature of 1400–1500 °C (Hanein et al. 2022). For standardization and large-scale development of LC3, its long-term performance is to be assessed (Rengaraju and Pillai 2023).

The declining use of river sand and the popularity of desalinated sea sand have increased the potential of chloride attack on concrete if used as aggregates in the system. Adoption of cements with greater resistance to chloride ion ingress is one of several strategies that have been suggested to address this issue. Limestone calcined clay cements are considered as binders in such cases because of their good cl^- binding capacity and resistance to chloride ingress (Wang et al. 2023). However, low-clinker binders are more vulnerable to a higher carbonation rate than the OPC system due to limited reserved alkalinity caused by consumption of portlandite in pozzolanic reactions (Shah et al. 2018). Steel's high rate of carbonation raises durability concerns that could limit its use in large-scale infrastructure projects by causing corrosion to begin earlier than anticipated. A lower pH causes the passive layer to break down, which is the primary cause of corrosion initiation in carbonating concrete. The low-clinker binders reduce the buffering capacity, which could accelerate the drop in pH (Hule and Pillai 2026). This chapter focuses on the mechanism of steel corrosion in limestone calcined cement concrete system and its performance when susceptible to chloride attack and carbonation. This chapter also explores the suitability of LC3 systems in structural applications by looking into their long-term durability and performance.

2 Fundamentals of Steel Corrosion in Concrete

Corrosion of steel is an electrochemical reaction involving the need for a cathode, an anode and an electrolyte. In reinforced concrete, the pore solution acts as the electrolyte. The mechanisms of corrosion initiation of embedded steel in concrete are discussed in the following sections.

2.1 Passive Film Formation

The pore solution in reinforced concrete is extremely alkaline (pH $\approx$ 12.5–13.5) and naturally provides embedded steel with a protective environment. The steel surface develops a thin, stable passive oxide film due to the high alkalinity, which serves as a barrier and significantly slows iron dissolution. The reinforcement can withstand prolonged corrosion as long as this passive layer is preserved. However, this protective layer becomes unstable and eventually disintegrates when aggressive substances like chlorides seep into the concrete or when carbonation reduces the alkalinity. Electrochemical reactions cause corrosion to start as soon as the steel is depassivated.

When iron oxidises at the anodic regions as shown in Eq. 1, electrons are released and move through the steel to the cathode region, where moisture reduces dissolved oxygen as given by Eq. 2.

$$\mathrm{Fe} \rightarrow \mathrm{Fe}^{2+} + 2\mathrm{e}^- \tag{1}$$

$$\mathrm{O}_2 + 2\mathrm{H}_2\mathrm{O} + 4\mathrm{e}^- \rightarrow 4\mathrm{OH}^- \tag{2}$$

Rust and other extensive corrosion products are created when the resultant ferrous ions combine with hydroxide ions to form iron hydroxides. These products create internal tensile stresses in the surrounding concrete because they take up a lot more space than the original steel. This eventually compromises the structural performance and service life of reinforced concrete elements by causing cracking, delamination, and spalling. Figure 1 shows the mechanism of corrosion of a steel bar embedded in concrete.

Figure 2 shows the schematic sketch of the sequence of steel corrosion in concrete. Corrosion of steel in concrete can be divided into two stages; corrosion initiation phase and corrosion propagation phase. During the initiation phase, aggressive substances such as carbon dioxide or chlorides penetrate the concrete cover and gradually reach the reinforcement level. During this stage, the passive oxide film protects the embedded steel from visible damage due to the highly alkaline environment of the concrete. The initiation phase ends and the passive film is destroyed when the chloride concentration reaches a critical threshold or when carbonation lowers the

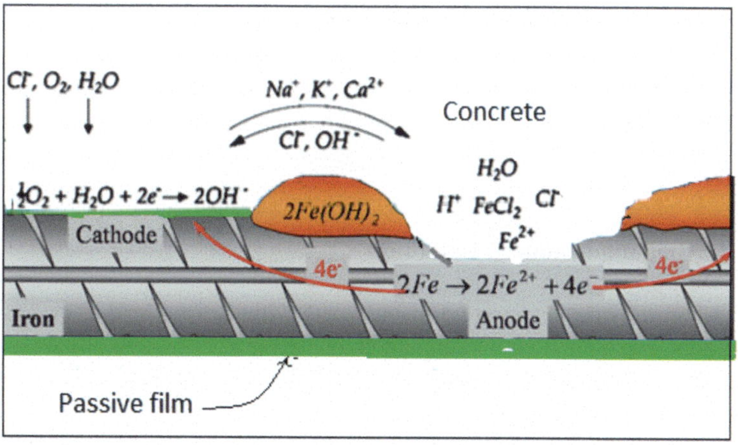

Fig. 1 Corrosion of a steel bar embedded in concrete (Gopu and Joseph 2022)

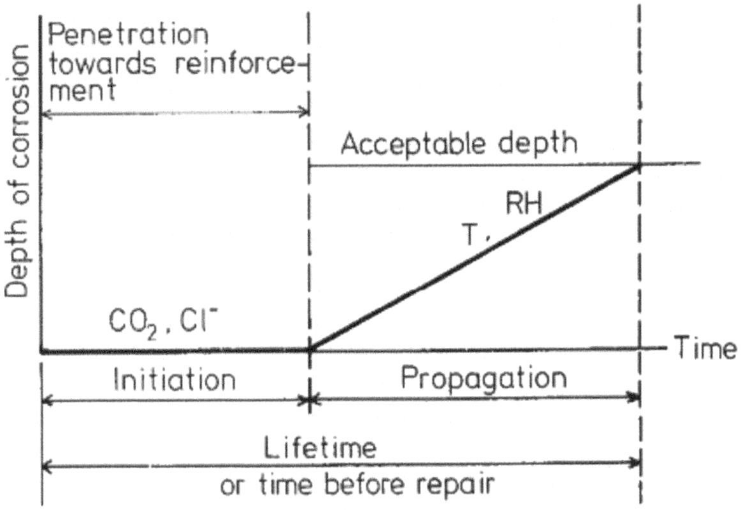

Fig. 2 Schematic sketch of corrosion sequence in concrete (Tuutti 1982)

pH to the point where depassivation takes place. The propagation phase begins as soon as corrosion begins. At this stage, extensive rust products are developed as a result of the active electrochemical corrosion of steel. The internal stresses that result in the concrete cover cracking, delaminating, and spalling gradually reduce the cross-sectional area of the reinforcement, putting the building's safety and structural integrity at risk. Thus, the total service life of a reinforced concrete structure is equal to the sum of its initiation and propagation periods.

2.2 *Corrosion Initiation Mechanisms*

Corrosion in steel is initiated by two mechanisms: carbonation-induced or chloride-induced. The mechanism and chemical reactions are explained in Sects. 2.2.1 and 2.2.2.

2.2.1 Carbonation-Induced Corrosion

When carbon dioxide in the atmosphere reacts with the portlandite or $Ca(OH)_2$ present in the concrete, it forms calcium carbonate ($CaCO_3$). Also, the CO_2 in the presence of moisture forms carbonic acid (H_2CO_3). The concrete's pH is lowered by these reactions and their byproducts, which depassivates the steel in the concrete. These chemical processes result in carbonates, bicoarbonates, and carbonic acid from carbon dioxide (as indicated by Eqs. 3, 4, and 5) causes steel in concrete to become carbonated. This process continues as more portalandite is reacted with carbonic acid.

$$Ca(OH)_2 + CO_2 \rightarrow CaCO_3 + H_2O \tag{3}$$

$$CO_2 + H_2O \rightarrow H_2CO_3 \tag{4}$$

$$H_2CO_3 + Ca(OH)_2 \rightarrow CaCO_3 + 2H_2O \tag{5}$$

The carbonation of concrete causes generalized corrosion and leads to the formation of microcell in corrosion in rebar (Chandru et al. 2021).

2.2.2 Chloride-Induced Corrosion

Chloride attack is one of the prominent reasons for corrosion of steel in concrete. When chloride ions come in contact with the surface of concrete in the presence of water and oxygen, the passive film of steel is locally destroyed. This often happens in coastal or marine regions where there is a chloride-rich environment. When the chloride ions penetrate into the concrete, it reacts with the steel to form ferrous chloride, which is a soluble complex. This leads to the formation of a 'pit' in the surface of rebar or embedded steel, which further causes the dissolution of more Fe^{2+} ions in the pore solution. These Fe^{2+} ions form ferric hydroxide upon reaction with hydroxyl ions and are finally converted to rust or hydrated ferric oxide upon oxidation reaction. The chloride attack on steel rebar can be explained using the Eqs. 6, 7, 8 and 9.

$$Fe^2 + + 2Cl_- \rightarrow FeCl_2 \tag{6}$$

$$Fe^{2+} + 2OH_- \rightarrow Fe(OH)2 \tag{7}$$

$$4Fe(OH)_2 + O_2 + 2H_2O \rightarrow 4Fe(OH)_3(\text{Ferric hydroxide}) \tag{8}$$

$$2Fe(OH)_3 \rightarrow Fe_2O_3.H_2O + 2H_2O\ (\text{Hydrated ferric oxide (rust)}) \tag{9}$$

In this process, Cl^- ions act as catalysts and stimulate further corrosion reactions as shown in Fig. 3. The steel corrosion products have a larger volume, up to 6 times more than the volume of the original steel, which is shown graphically in Fig. 4.

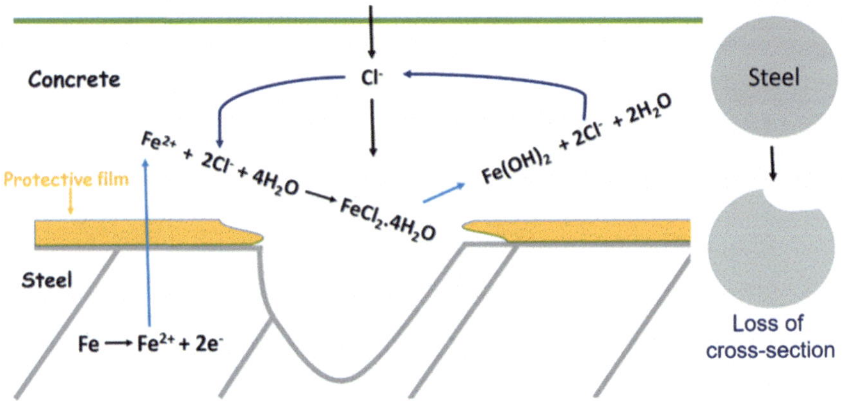

Fig. 3 Catalytic role of chloride ions in steel corrosion (Huyen Vu et al. 2022)

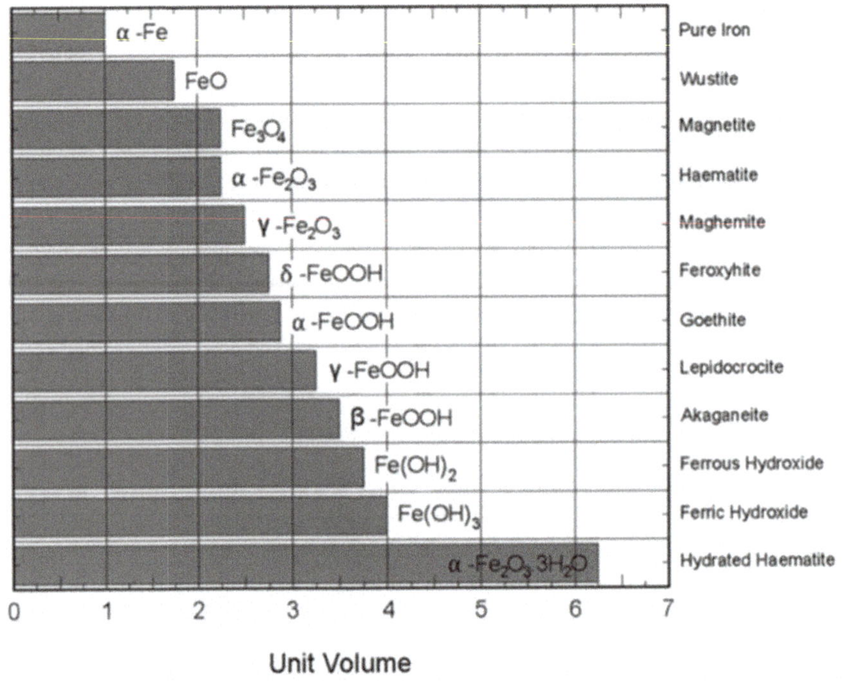

Fig. 4 Volume of corrosion products compared to the volume of original steel (Jaffer and Hansson 2009)

3 Microstructural Refinement of LC3 Concrete

LC3 systems have the potential to be used in high-performance concrete without any ultra-fine SCMs, due to their microstructural refinement. When compared to OPC and Fly ash based concrete mixes, LC3 mixes show a different and denser microstructure (Dhandapani et al. 2018). Along with matrix densification filler effect is observed in LC3 system (Ejbouh et al. 2023; Ahmed et al. 2023; Sui et al. 2023). Due to the presence of more reactive silica and alumina in LC3, both outer and inner C–S–H formation caused noticeable composition changes as shown in Fig. 5.

The microstructure development of LC3 is assessed using methods like Mercury intrusion porosimetry (MIP) and Electrical impedance spectroscopy (EIS). In comparison to OPC and fly ash blended cement, the LC3 binder reaches a highly refined pore structure by 7 days. Due to the high reactivity of calcined clay, X-ray diffraction results also shows higher $Ca(OH)_2$ consumption in the early age.

The presence of monocarboaluminate and hemicarboaluminate, indicated the progressive microstructure development (Dhandapani and Santhanam 2017). Scanning Electron Microscopy also proves to be useful for assessing the microstructure by identifying pores and hudaration products (Sui et al. 2023). Figure 6 shows the change in pore sizes of the OPC, fly ash blended system and LC3 system. These results align with another study by (Bernal et al. 2021), where methods like MIP and powder diffraction were employed to investigate the pore connectivity and phase development in the LC3 system, respectively. A decreased pore connectivity between 28 and 60 days of hydration was observed. On comparing the pore volume, LC3 system shows a 17% lesser value than OPC system, which also points to lesser pore connectivity (Sui et al. 2023).

Furthermore, the effects of elevated temperatures on the strength and pore structure of LC3 were studied by (Lin et al. 2021), and it was found that unhydrated binder particles continued to hydrate from 20–300 °C. The total pore volume didn't undergo

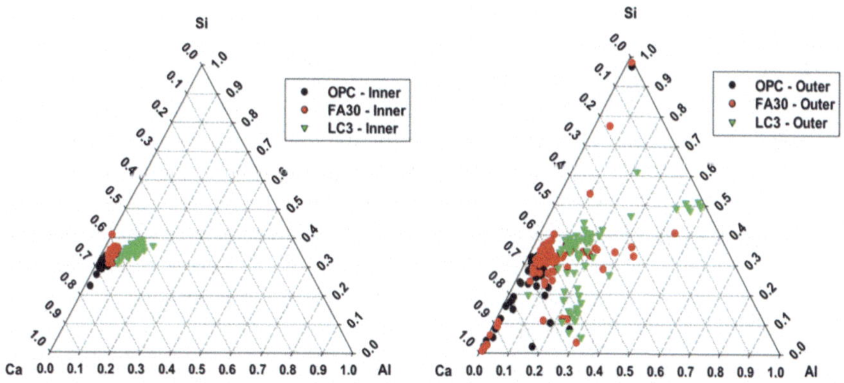

Fig. 5 Composition of inner and outer CSH in the OPC, Fly ash blended, and LC3 cementitious systems at 150 days. (Dhandapani et al. 2018)

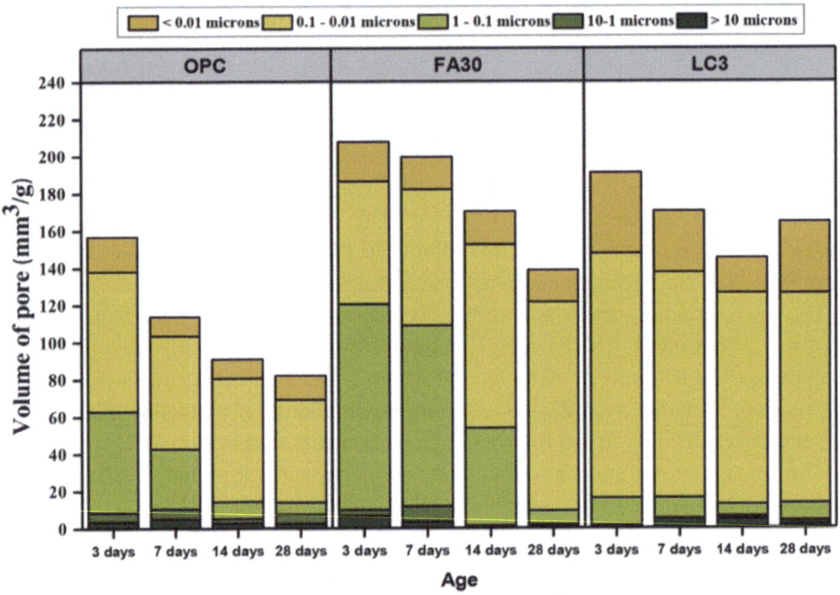

Fig. 6 Change in the pore sizes of OPC, fly ash blended and LC3 system (Dhandapani and Santhanam 2017)

much change, but the pores became smaller and more refined. This microstructure refinement and reduced pore connectivity increased the strength of the mix. But, at high temperatures, this behaviour started to reverse. C–S–H gel and portlandite started decomposing at temperature ranges of 400–600 °C. When the temperature was further elevated to 900 °C, severe phase breakdown occurred, resulting in high porosity and strength loss.

4 Transport Mechanisms in LC3 Concrete

The transport parameters in concrete are greatly influenced by the pore structure and network of the hydrated cement matrix, which is identified based on the electrical tortuosity (Yang et al. 2017). It shows the state of complex pore network in the microstructure of cement pastes (Dhandapani and Santhanam 2020) and is defined as the ratio of the length of the transport path to the shortest distance between the start and end points of the path (Yang et al. 2017).Since it is difficult to measure the tortuosity directly, it is taken as a function of pore connectivity obtained from electrical conductivity measurements. LC3 systems attain significantly higher tortuosity state at an early age in comparison to fly ash blended systems (Dhandapani and Santhanam 2020).

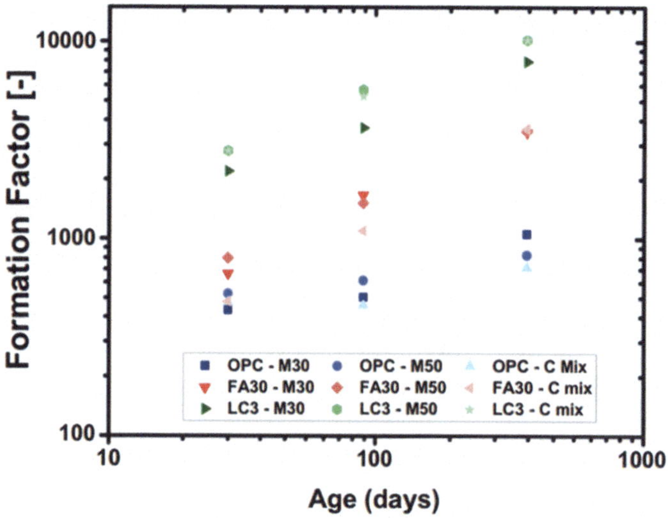

Fig. 7 Formation factor of different concrete mixes (Dhandapani and Santhanam 2020)

Another factor that affects the pore structure of concrete systems is the formation factor, also known as the formation resistivity factor. It is defined as the ratio of the conductivity of the pore solution to the conductivity of the bulk concrete (Dhandapani and Santhanam 2020). It shows how much the microstructure restricts ionic movement compared to a free pore solution. A higher formation factor indicates discontinuous or refined pores, meaning the ions move slowly, making the system more resistive and denser. When comparing the formation factors of OPC, fly ash blended and LC3 systems, LC3 showed the highest formation factor at all ages, as depicted in Fig. 7.

The other factors affecting the transport properties in the concrete system apart from pore structure are pore solution and binding capacity. Binding capacity is the factor responsible for fixing the chloride ions in the hydration products. A higher binding capacity indicates slower chloride transport (Sui et al. 2019). Calcined clay system tends to have a higher binding capacity in comparison to OPC, slag and fly ash-based systems as shown in Fig. 8.

5 Service Life Parameters of LC3 Concrete

A reinforced concrete system's service life is often reduced by corrosion brought on by chloride attack, and the corrosion initiation phase is marked as the end of the service life. The duration of this phase is determined by the chloride diffusion coefficient (Dcl), ageing coefficient for Dcl (m), surface chloride concentration (Cl_s), cover depth (d), and chloride threshold (Cl_{th}).

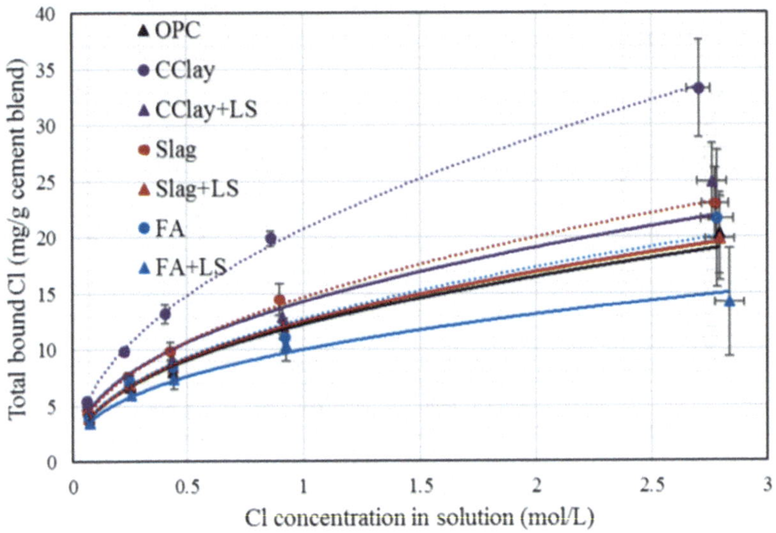

Fig. 8 Total bound chlorine for different binder systems (Sui et al. 2019)

When the D_{cl}, m and Cl_{th} related to seven concrete mixes (four M35 and three M50) with OPC, OPC + PFA (pulverised fuel ash) and limestone-calcined clay cement (LC3) were studied, the OPC mixes showed higher rates of chloride ingress compared to PFA and LC3 concretes (Pillai et al. 2019). This is explained by the less porous microstructure and high ionic resistance of the binder phases of the PFA and LC3 concretes. The near-surface porosity of concrete is responsible for the variations in surface chloride concentration. Surface chloride concentrations were lower in the LC3 and PFA concretes due to their low porosity and smaller pore sizes. Table 1 shows the results of the tests for different mixes.

From these results, it can be interpreted that OPC mixes showed very high diffusion values, whereas PFA and LC3 showed much lower values, where LC3 shows an 80–90% reduction in diffusion compared to OPC. In regard to electrical conductivity, LC3 mixes showed extremely low values in the range of 0.14–0.18 mS/m. This shows low ionic movement and a lesser risk of reinforcement corrosion. Ageing coefficient of LC3 mixes showed a higher value in the range of 0.5–0.55, indicating a reduction in diffusion, which may be due to continued pozzolanic reactions in the system. Although LC3 mixes showed a lower chloride threshold value in the range of 0.02–0.03%, their diffusion rate is very low; chloride takes much longer to reach the reinforcement. So, it can be concluded that LC3 mixes have a longer service life compared to OPC mixes. These findings are in alignment with previous studies (Dhandapani et al. 2018). The surface chloride resistivity of OPC, fly ash-based blended cement (FA30), and LC3 were measured, the results of which are shown in

Table 1 Different concrete mixes and their relevant parameters (Pillai et al. 2019)

Ingredients/properties		OPC-1	PFA-1	LC3-1S	LC3-1P
Binder type		OPC	OPC + PFA	LC3	LnC3
Binder content (kg/m^3)		310	360	310	360
Water-binder ratio		0.5	0.45	0.5	0.45
Diffusion coefficient at 365 Days ($\times 10^{-12}$ m^2/s)	μ	18.7	3.5	3.3	1.3
	cov	0.34	0.2	0.08	0.12
Electrical conductivity (mS/m)	28 days	11.87	7.79	0.65	0.51
	90 days	10.19	3.4	0.39	0.27
	365 days	4.87	1.03	0.18	0.14
Ageing coefficient (m)	μ	0.26	0.7	0.5	0.53
	cov	0.3	0.14	0.27	0.14
Chloride threshold (Cl_{th}) (% bwoc)	μ	0.06	0.05	0.02	0.03
	cov	0.28	0.32	0.39	0.39

Figs. 9 and 10. The findings demonstrated that LC3 systems had the highest resistivity, suggesting that early pore refinement caused the resistivity to rise at early ages.

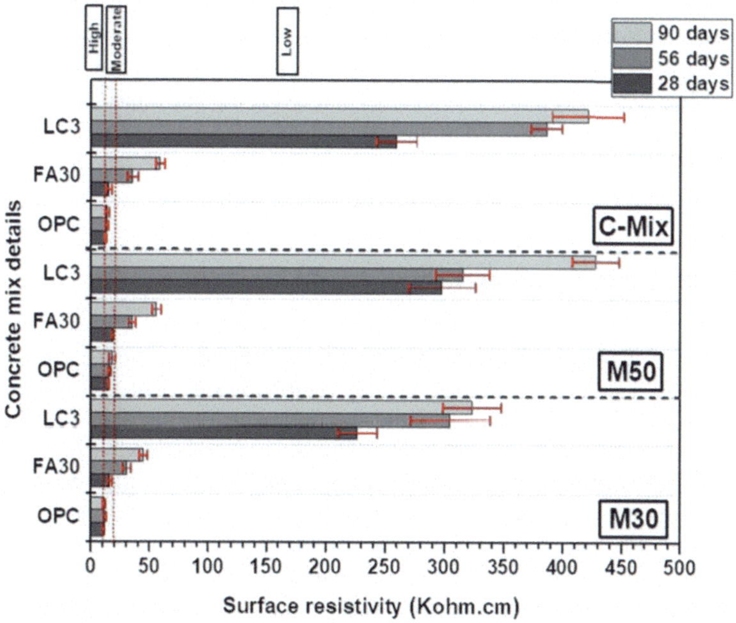

Fig. 9 Surface resistivity of the different concretes at 28 days, 56 days and 90 days (Dhandapani et al. 2018)

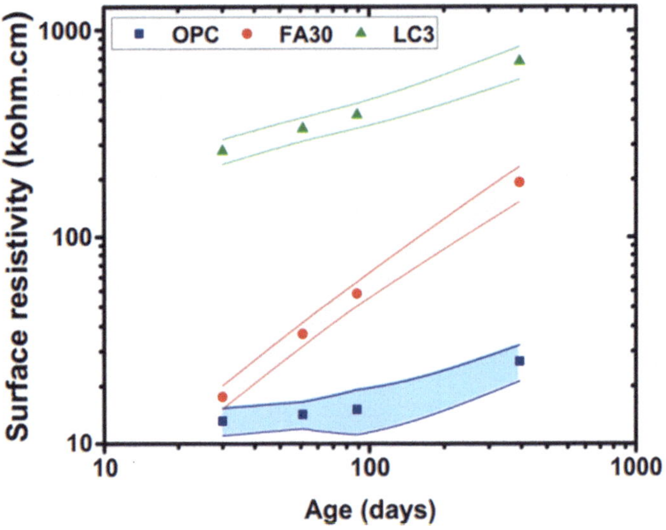

Fig. 10 Development of surface resistivity of the different concretes (Dhandapani and Santhanam 2020)

LC3 system performs better in against chloride ingress than OPC system, but are still susceptible to pitting corrosion. This can be controlled using corrosion inhibitors (Rengaraju and Pillai 2023).

In the LC3 system, higher clinker replacement levels lead to more refined pores, higher surface resistivity, and enhanced resistance to sulphate attack. This is due to an increase in hemicarbonate and monocarbonate phases (Arbab et al. 2025; Ejbouh et al. 2023).

6 Resistance to Carbonation of LC3 Concrete

Carbonation is a physicochemical process that affects the durability of concrete systems. Many studies show that concretes with SCMs tend to be more susceptible to carbonation due to the pozzolanic reaction, which reduces the CO_2 binding capacity of the concrete (Hule and Pillai 2026; Shah et al. 2018). Factors affecting carbonation in concrete systems include relative humidity (RH), pH, resistivity and oxygen concentration at the steel—concrete interface. On comparing these properties of OPC, Fly ash blended, and LC3 systems, exposed to 3% CO2 to induce carbonation, it was found that if the relative humidity was above 75%, it increased the corrosion rate. The variation in pH of these three systems is as shown in Fig. 11. The reduction in pH of the blended systems is due to the pozzolanic reaction and a subsequent reduction in portlandite content. Figure 12 shows the changes in the microstructure of these systems before and after carbonation. Before carbonation,

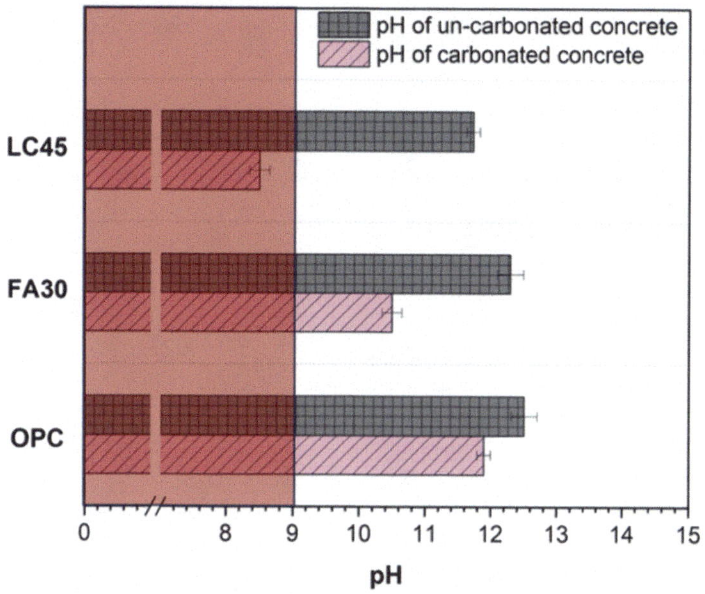

Fig. 11 The pH of carbonated and uncarbonated mortar powder solution using modified ASTM C25–19 (Hule and Pillai 2026)

low clinker binders showed lower transport due to narrower pores and longer paths for ions to travel. As a result the ITZ (interfacial transition zone) is improved. After carbonation, in OPC, calcium carbonate is formed due to the availability of portlandite. This fills the pore, resulting in a denser ITZ. But in low clinker binders, due to lesser availability of portlandite, carbonation attacks C–S–H gel leading to its decalcification. This causes pore coarsening and redistribution which increases the pore connectivity leading to an increase in transport properties (Table 2).

By taking into account the extent of CO_2 capture and storage, this LC3 system drawback can be mitigated. Because LC3 systems absorb relatively more carbon, they can be regarded as a promising technique for carbon capture and CO_2 storage. The disadvantage of carbonation in LC3 systems can be turned into a benefit in the context of CO2 capture with further research on its volume stability, durability, and chemical erosion. As a result, carbon sequestration is a potential future in the case of LC3 concrete (Cho et al. 2026; Hanifa et al. 2023).

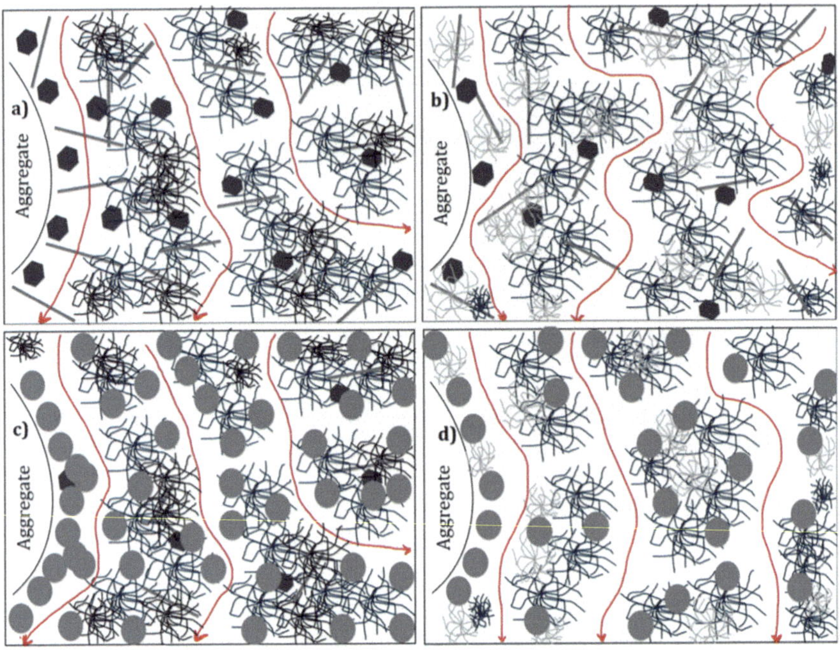

Fig. 12 Schematic diagram of microstructure in **a** PC–NC, **b** Blended cement—NC, **c** PC–C, and **d** Blended cement—C. (Hexagons–Calcium hydroxide, rods—ettringite, spheres—calcium carbonate, and foils-like structure—C–S–H or decalcified C–S–H) (Dudi and Bishnoi 2025)

Table 2 Parameters before and after carbonation (Shah et al. 2018)

Mix	OPC	PPC	LC3
% Change in porosity (MIP)	1.54	− 0.32	− 1.22
% Change in solid volume (Complete carbonation – thermodynamic calculations)	− 12.22	− 11.11	− 13.90
% Change in solid volume (on the basis of experimentally measured degrees of carbonation)	1.55	− 2.95	− 6.63
% Change in solid volume (on the basis of experimentally measured degrees of carbonation and considering additional hydration)	5.74	− 0.50	− 4.11
Amount of carbon dioxide bound in accelerated carbonated samples per 100 g of powder	12.13	18.41	12.95
The degree of carbonation (%)	18	40	39

7 Electrochemical Assessment of Steel Embedded in Reinforced LC3 Concrete

The various test methods for the electrochemical assessment and surface characterization of steel in reinforced limestone calcined clay cement concrete are as shown in Table 3. The LC3 system demonstrated a lower corrosion rate in the passive stage, indicating a higher polarisation resistance and a lower corrosion current, according to the experimental results of these tests. Reactive alumina causes the formation of AFm phases, which enhance chloride binding and lower the concentration of free chloride at the steel surface (Li et al., n.d.).The passive film that forms in the LC3 pore solution is thinner than that of the OPC system, measuring 4.9 nm for LC3 and 5.0 nm for OPC. There is more calcium incorporation in the OPC passive film. The corrosion rate in the passive stage of LC3 is three times lower than that of OPC, despite the thinner passive film and lower pH. Depassivation is delayed and the chloride threshold is 4.6 times higher.

Table 3 Electrochemical test and surface characterization methods and evaluated parameters (Li et al. 2022; Nguyen and Castel 2020; You et al. 2025)

Test method	Full form	Main parameter measured	Corrosion information obtained
OCP	Open circuit potential	Corrosion potential (E_{corr})	More negative potential; higher probability of corrosion initiation
LPR	Linear polarization resistance	Polarization resistance (R_p), Corrosion current density (I_{corr})	Lower I_{corr}; lower corrosion rate
EIS	Electrochemical impedance spectroscopy	Charge transfer resistance (R_{ct}), Film resistance (R_f), Capacitance	Higher R_{ct}; better corrosion resistance
Electrochemical Monitoring after Exposing to NaCl	–	Chloride threshold (Cl^-/OH^- ratio)	Higher threshold; better chloride tolerance
XPS	X-Ray photoelectron spectroscopy	Chemical composition, Fe^{2+}/Fe^{3+} ratio	–
AFM	Atomic force microscopy	Surface roughness, morphology	–
SEM–EDS	Scanning electron microscopy coupled with energy dispersive spectroscopy	Elemental distribution	–
XRD	X-Ray diffraction	Crystalline phases (AFm, Friedel's salt)	–

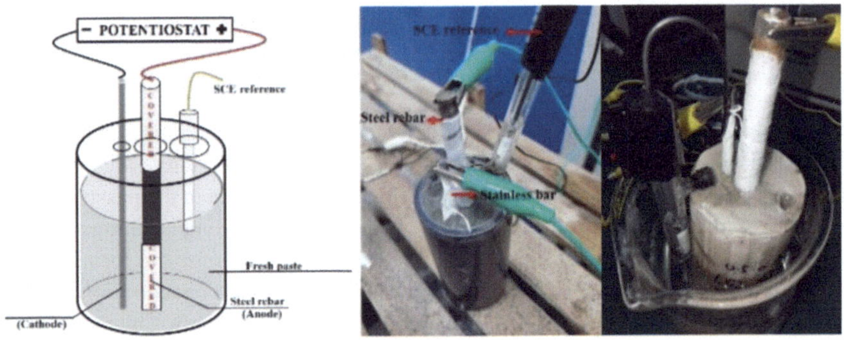

Fig. 13 Three-electrode system for electrochemical measurements (Li et al. 2022)

On the steel surface of LC3, a 3 μm-thick layer of AFm (hemicarbonate/monocarbonate) rich in aluminium is discovered. Despite the LC3 system's thinner passive film, this turns out to be the cause of its enhanced chloride resistance. The AFm layer can bind chloride ions and lower the concentration of free chloride at the steel surface due to its layered double hydroxide structure (You et al. 2025). The typical experimental setup for electrochemical measurements is shown in Fig. 13.

Steel embedded in LC3 systems exhibited a quicker decrease in corrosion potential and an increase in corrosion current density when subjected to coupled carbonation and chloride attack. There was a decrease in both the charge transfer resistance and the polarisation resistance. In contrast to single exposure conditions, the corrosion starts earlier. The AFm phases that bind chlorides break down during carbonation in a combined chloride and carbonation attack. As a result, previously bound chlorides are released back into the pore solution. The concentration of free chloride close to the steel surface rises as a result, raising the possibility of depassivation (Hay and Celik 2024). The corrosion mechanism under coupled or combined chloride and carbonation attack is shown in Fig. 14.

7.1 Long-Term Durability and Field Performance

Assessment of the long-term durability of LC3 concrete is difficult, as many of the projects are in the pilot stage. But accelerated tests can be a tool to predict its behaviour in the long term. When comparing the performance of OPC and LC3 systems, embedded steel in LC3 systems shows more localized corrosion, whereas the OPC system shows a more uniform corrosion. Corrosion ratio is a factor that helps in predicting the performance of the systems. Corrosion ratio is the percentage loss of steel cross-section or mass due to corrosion compared to its original condition. It tells how much steel has deteriorated. When the corrosion ratio is increased, the mechanical properties are affected. An increase in corrosion ratio from 5 to 10%, increases the reduction in the flexural strength from 11.5% to 27.2%. Similarly, a

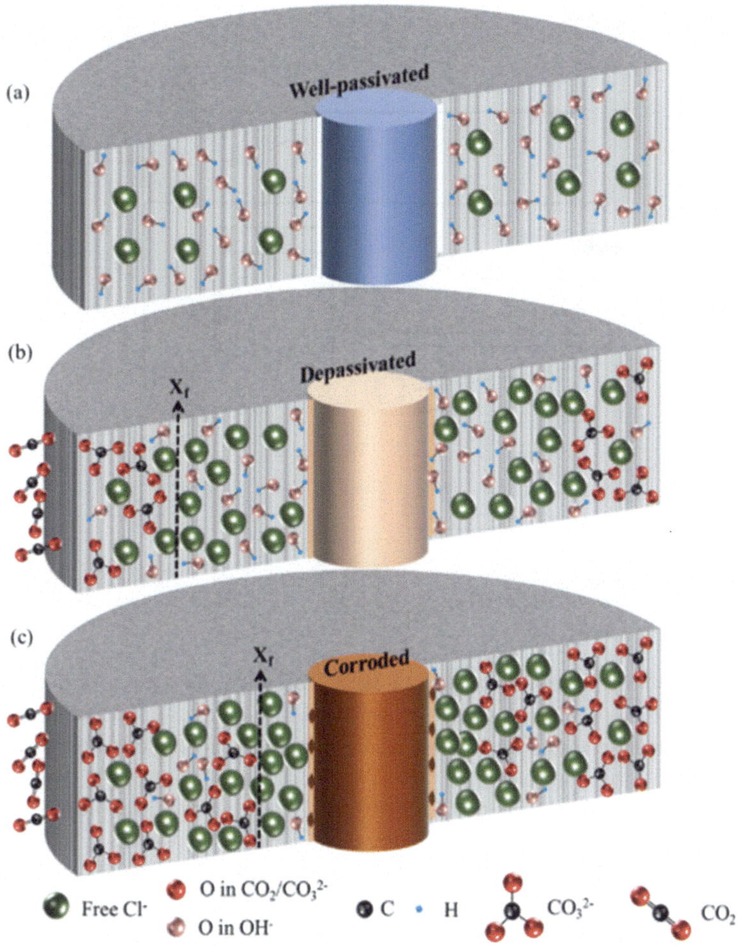

Fig. 14 The corrosion mechanism under combined chloride and carbonation attack. (Hay and Celik 2024)

reduction in ductility is also observed from 13.4% to 22.8% (Huang et al. 2026). So, 10% can be considered as acritical corrosion threshold. Structural performance deteriorates rapidly after this point.

To predict the long-term performance of a reinforced LC3 structural system, experimental tests along with mathematical modelling can be done. To induce environmental parameters on the system axial loading with chloride attack can be tested. Figure 15. Shows a typical test setup for applying axial load with chloride-induced corrosion on concrete piers. It mimics the marine environment exposure to the system. The load deflection behaviour of a reinforced concrete system is analysed to study the performance. The initial stiffness of the member is not significantly affected by corrosion, but the yield deflection and ultimate deflection are reduced significantly.

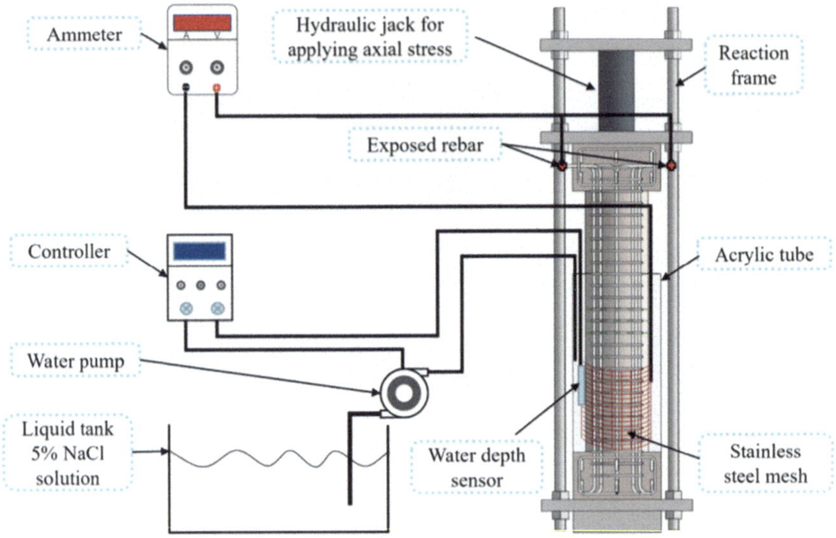

Fig. 15 Schematic diagram of setup for applying axial loading on concrete piers with corrosion induced by chloride wet-dry cycles (Huang et al. 2026)

When the critical corrosion threshold of 10% is attained, the system showed asymmetric deflection, premature rebar fracture and brittle failure. The bond strength between the steel and concrete also reduced. The axial load played a significant role in reducing the corrosion ratio. When the axial load was increased from 0 to 40%, the corrosion ratio decreased from 6.8% to 2.8%. This may be due to suppressed microcrack development due to axial compression. So, axial load helps to mitigate corrosion damage or severity to an extent. Figure 16a, b shows the effects of different rates of axial loading and different type of concrete on actual corrosion rates. To test the long term performance theoretical model was developed that matched the experimental results.

8 Conclusions

Limestone calcined clay cement concrete is a promising material for widespread applications. This chapter was mainly focused on the corrosion of embedded steel in LC3 system. From the results of various experimental studies and models, the following conclusions are arrived at;

- Even without ultra-fine supplementary cementitious materials, LC3 systems exhibit great promise for high-performance concrete due to their dense and refined microstructure. Compared to OPC and fly ash blends, LC3 exhibits better matrix

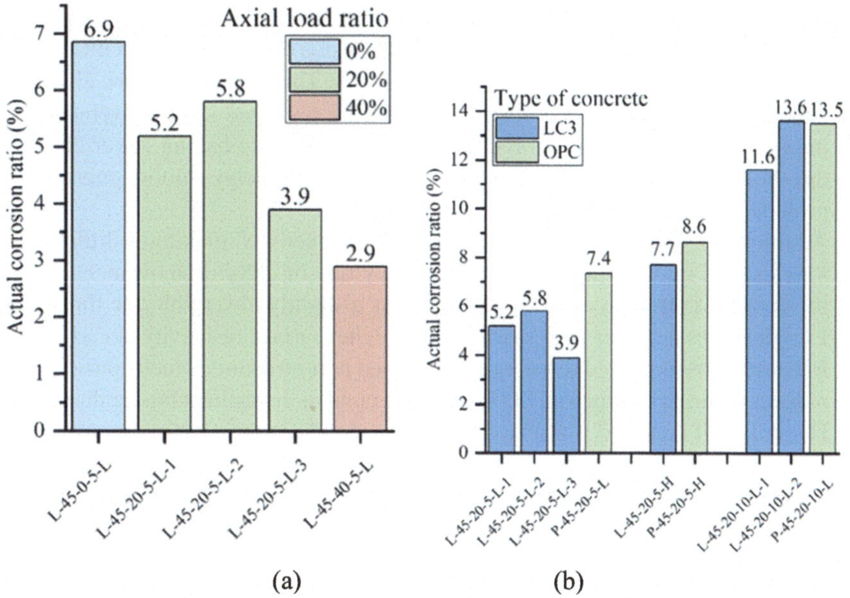

Fig. 16 **a** Effects of axial load ratio on actual corrosion ratio (Huang et al. 2026); **b** Effects of concrete type on actual corrosion ratio (Huang et al. 2026)

densification through filler effects and the creation of additional hydration products. Because calcined clay is more reactive, it changes the composition of inner and outer C–S–H and consumes more $Ca(OH)_2$. Microstructural analyses using MIP, EIS, XRD, and SEM confirm reduced pore volume, decreased pore connectivity, and the formation of stable carboaluminate phases. These characteristics help in improving the mechanical performance and durability at both early and later ages. On the other hand, the breakdown of hydration products at very high temperatures leads to an increase in porosity and a decrease in strength.

- Pore structure of concrete, especially its tortuosity and pore connectivity, has a significant influence on its transport properties. Ionic movement within the cement matrix is greatly influenced by electrical tortuosity. Compared to fly ash blended systems, LC3 systems show greater tortuosity at early ages, suggesting a more complex and irregular pore structure. Similarly, the formation factor of LC3 is higher than that of fly ash and OPC systems, suggesting that reduced pore connectivity is the cause of increased resistance to ionic transport. Transport behaviour is further regulated by pore solution chemistry and chloride binding capacity in addition to pore structure. The improved durability performance of LC3 concrete is a result of the slower chloride ingress caused by the increased chloride binding capacity of calcined clay-based systems.
- The service life of reinforced concrete is largely controlled by chloride-induced corrosion, with factors like the diffusion coefficient, ageing coefficient, surface chloride concentration, cover depth, and chloride threshold being crucial during

the initiation phase. Experimental findings show that LC3 and Fly ash-based concretes have significantly lower chloride diffusion coefficients than OPC mixes; LC3 achieves an 80–90% reduction in diffusion. The significantly low electrical conductivity values of LC3 further confirm the restricted ionic movement and improved resistance to reinforcement corrosion. Despite having lower chloride threshold values, lower diffusion rates of LC3 systems delay chloride ingress and prolong the time it takes for corrosion to start.

- Additionally, LC3 has a very low electrical conductivity, making it difficult for ions to pass through the concrete. The slow rate of chloride movement delays the onset of corrosion even though LC3 has a slightly lower chloride threshold. LC3 develops a denser pore structure and higher surface resistivity at early ages. Because of its improved resistance and refined pore structure, which considerably reduces chloride infiltration, LC3 is an all-around more resilient binder than OPC.
- Durability of concrete is affected by carbonation because it lowers its alkalinity and raises the possibility of steel corrosion. Because they contain less portlandite, which typically helps in resisting pH reduction, blended systems like LC3 are typically more susceptible to carbonation. LC3 has a refined pore structure with poorer transport characteristics before carbonation, but pore coarsening and increased connectivity may result from the decalcification of C–S–H gel following carbonation. When exposed to carbonation, this alteration might slightly decrease durability and increase permeability. However, LC3 has the ability to bind a significant amount of CO^2, which may be advantageous for carbon capture and storage. Therefore, with additional research and optimisation, LC3 systems offer promising opportunities for sustainable and carbon-sequestering concrete, even though carbonation presents a durability challenge.
- The loss of steel cross-section and bond strength is the primary way that corrosion directly affects the structural performance of reinforced LC3 concrete. Significant decreases in flexural strength and ductility are seen as the corrosion ratio rises, particularly above the critical threshold of roughly 10%. Members may exhibit brittle failure, asymmetric deflection, and early rebar fracture at this point. The yield and final deflections decrease significantly as corrosion advances, even though the initial stiffness is not significantly impacted. Therefore, preventing excessive steel deterioration and controlling corrosion levels are critical to the structural safety of LC3 reinforced members.

Acknowledgements The authors gratefully acknowledge the financial support provided by the Anusandhan National Research Foundation (ANRF), Government of India, under Project Sanction No. ANRF/ PDF/2022/001248, titled "Carbonation-induced corrosion and service life of steel-concrete systems with limestone calcined clay cement (LC3) and corrosion inhibitors."

References

Ahmed AH, Nune S, Liebscher M, Köberle T, Willomitzer A, Noack I, Butler M, Mechtcherine V (2023) Exploring the role of dilutive effects on microstructural development and hydration kinetics of limestone calcined clay cement (LC3) made of low-grade raw materials. J Cleaner Prod 428. https://doi.org/10.1016/j.jclepro.2023.139438

Arbab F, Tarighat A, Shayanfar MA (2025) Evaluation of electrical resistivity, hydration, and durability against sulfate attack of limestone calcined clay cement (LC3) with various kaolinite content at high-replacement levels. Constr Build Mater 469. https://doi.org/10.1016/j.conbuildmat.2025.140505

Bernal IMR, Shirani S, Cuesta A, Santacruz I, Aranda MAG (2021) Phase and microstructure evolutions in LC3 binders by multi-technique approach including synchrotron microtomography. Constr Build Mater 300. https://doi.org/10.1016/j.conbuildmat.2021.124054

Chandru P, Karthikeyan J, Natarajan C (2021) Techniques to assess the corrosion resistance and corrosion rate of the steel embedded in concrete, pp 33–54. https://doi.org/10.1007/978-3-030-71233-4_3

Cho A, Seo J, Lee HK (2026) Reaction products and reactivity of carbonation-cured limestone calcined clay cement (LC3). Mater Struct/Mater Constr 59(1). https://doi.org/10.1617/s11527-025-02901-9

Dhandapani Y, Sakthivel T, Santhanam M, Gettu R, Pillai RG (2018) Mechanical properties and durability performance of concretes with limestone calcined clay cement (LC3). Cem Concr Res 107:136–151. https://doi.org/10.1016/j.cemconres.2018.02.005

Dhandapani Y, Santhanam M (2017) Assessment of pore structure evolution in the limestone calcined clay cementitious system and its implications for performance. Cement Concr Compos 84:36–47. https://doi.org/10.1016/j.cemconcomp.2017.08.012

Dhandapani Y, Santhanam M (2020) Investigation on the microstructure-related characteristics to elucidate performance of composite cement with limestone-calcined clay combination. Cem Concr Res 129. https://doi.org/10.1016/j.cemconres.2019.105959

Dudi L, & Bishnoi, S. (2025). Revealing the relationships between transport properties and microstructure characteristics in low-clinker binders before and after carbonation. Cement Concrete Res 198. https://doi.org/10.1016/j.cemconres.2025.108006

Ejbouh A, Ech-chebab A, Hassi S, Galai M, Benqlilou H, Ebn Touhami M (2023) Durability assessment of LC3-based reinforced concrete under combined chloride-sulfate environment via the EIS technique. Constr Build Mater 366. https://doi.org/10.1016/j.conbuildmat.2022.130194

Ez-zaki H, Marangu JM, Bellotto M, Dalconi MC, Artioli G, Valentini L (2021) A fresh view on limestone calcined clay cement (LC3) pastes. Materials 14(11). https://doi.org/10.3390/ma14113037

Gopu GN, Joseph SA (2022) Corrosion behavior of fiber-reinforced concrete—A review. In: Fibers, vol 10, no 5. MDPI. https://doi.org/10.3390/fib10050038

Hanein T, Thienel KC, Zunino F, Marsh A, TM, Maier M, Wang B, Canut M, Juenger MCG, Ben Haha M, Avet F, Parashar A, Al-Jaberi LA, Almenares-Reyes RS, Alujas-Diaz A, Scrivener KL, Bernal SA, Provis JL, Sui T, Bishnoi S, Martirena-Hernández F (2022) Clay calcination technology: state-of-the-art review by the RILEM TC 282-CCL. In: Materials and structures/materiaux et constructions (vol 55, no 1). Springer Science and Business Media B.V. https://doi.org/10.1617/s11527-021-01807-6

Hanifa M, Agarwal R, Sharma U, Thapliyal PC, Singh LP (2023) A review on CO_2 capture and sequestration in the construction industry: emerging approaches and commercialised technologies. J CO_2 Utilization 67:102292

Hay R, Celik K (2024). Comparative performance of OPC and LC3-based composites under combined action of carbonation and chloride exposure. Constr Build Mater 429. https://doi.org/10.1016/j.conbuildmat.2024.136428

Huang Z, Zhao X, Xie Y, Chen L, Zhou Y (2026) Flexural performance of limestone calcined clay cement (LC3) concrete bridge piers under axial-lateral loading and chloride-induced corrosion. Eng Struct 354:122383. https://doi.org/10.1016/j.engstruct.2026.122383

Hule U, Pillai RG (2026) Corrosion of steel in concretes with fly ash and limestone calcined clay exposed to carbon dioxide. Constr Build Mater 515. https://doi.org/10.1016/j.conbuildmat.2026.145620

Huyen Vu T, Dang LC, Kang G, Sirivivatnanon V (2022) Chloride induced corrosion of steel reinforcement in alkali activated slag concretes: a critical review. Case Stud Constr Mater 16. https://doi.org/10.1016/j.cscm.2022.e01112

Jaffer SJ, Hansson CM (2009) Chloride-induced corrosion products of steel in cracked-concrete subjected to different loading conditions. Cement Concr Res 39(2):116–125

Li W, Xu Z, Wang Y, Wang X, Xing F (2022) Corrosion Performance of Embedded Steel Bar in Cl--contaminated Limestone Calcined Clay Cement (LC3) at initial stage of hydration. In: International conference on durability of concrete structures, p 6. https://docs.lib.purdue.edu/icdcs/2022/partii/6

Lin RS, Han Y, Wang XY (2021) Macro–meso–micro experimental studies of calcined clay limestone cement (LC3) paste subjected to elevated temperature. Cement and Concrete Composites, 116. https://doi.org/10.1016/j.cemconcomp.2020.103871

Nguyen QD, Castel A (2020) Reinforcement corrosion in limestone flash calcined clay cement-based concrete. Cement and Concrete Res 132. https://doi.org/10.1016/j.cemconres.2020.106051

Pillai RG, Gettu R, Santhanam M, Rengaraju S, Dhandapani Y, Rathnarajan S, Basavaraj AS (2019) Service life and life cycle assessment of reinforced concrete systems with limestone calcined clay cement (LC3). Cem Concr Res 118:111–119. https://doi.org/10.1016/j.cemconres.2018.11.019

Rengaraju S, Pillai G, R (2023) Long-term corrosion performance and monitoring for service life estimation of LC3 concrete systems. J Sustain Cem-Based Mater 12(12):1592–1603. https://doi.org/10.1080/21650373.2023.2246068

Scrivener K, Martirena F, Bishnoi S, Maity S (2018) Calcined clay limestone cements (LC3). In: Cement and concrete research, vol 114. Elsevier Ltd., pp 49–56 https://doi.org/10.1016/j.cemconres.2017.08.017

Shah V, Scrivener K, Bhattacharjee B, Bishnoi S (2018) Changes in microstructure characteristics of cement paste on carbonation. Cem Concr Res 109:184–197. https://doi.org/10.1016/j.cemconres.2018.04.016

Sui H, Hou P, Liu Y, Sagoe-Crentsil K, Basquiroto de Souza F, Duan W (2023) Limestone calcined clay cement: mechanical properties, crystallography, and microstructure development. J Sustain Cem-Based Mater 12(4):427–440. https://doi.org/10.1080/21650373.2022.2074911

Sui S, Georget F, Maraghechi H, Sun W, Scrivener K (2019) Towards a generic approach to durability: factors affecting chloride transport in binary and ternary cementitious materials. Cement Concrete Res 124. https://doi.org/10.1016/j.cemconres.2019.105783

Tuutti K (1982) Corrosion of steel in concrete. Swedish Cement & Concrete Inst. Report Fo4

Vavpetič P (2008) Corrosion in concrete steel

Wang Y, Xu Z, Mi T, Yu J, Xing F, Li W (2023) Corrosion mechanism of reinforcement in LC3 cement pastes under coupled carbonation and chloride attack. Cement and Concrete Compos 140. https://doi.org/10.1016/j.cemconcomp.2023.105080

Yang P, Sant G, Neithalath N (2017) A refined, self-consistent poisson-nernst-planck (PNP) model for electrically induced transport of multiple ionic species through concrete. Cement Concr Compos 82:80–94. https://doi.org/10.1016/j.cemconcomp.2017.05.015

You Q, Zhang Z, Zheng Y, Hu Y, Liu Q, Geng G (2025) Steel passivation in LC3 pore solution influenced by AFm deposition. Construc Build Mater, 505. https://doi.org/10.1016/j.conbuildmat.2025.144740

Corrosion Mechanisms, Inspection Methods and Repair Strategies for Anchorage Regions in Grouted, Post-tensioned Concrete Bridges

Karthikeyan Manickam and Chandru Pichaimuthu

Abstract Grouted, Post-tensioned (PTd) concrete systems are commonly used to construct bridges with an anticipated corrosion-free design service life of 100+ years. The anchorage region serves as the backbone of a PTd concrete system, carrying the load and ensuring the structural integrity. However, premature corrosion and tendon failures in the anchorage region have been observed in many PTd concrete bridges worldwide. This chapter presents a comprehensive review of the corrosion mechanisms in the anchorage region of grouted, PTd concrete systems. The mechanisms include corrosion due to chlorides, sulphates, carbon dioxide, oxygen, galvanic interactions, grout deficiencies, and inadequate anchorage protection systems. The review aims to help readers understand the technical concepts, significance, and implications of these corrosion mechanisms. Additionally, a review of both non-destructive and semi-destructive methods for assessing the condition of the anchorage region is included, enabling readers to select suitable inspection techniques for various site conditions. Finally, the paper reviews various corrosion control and prevention techniques developed to extend the service life of PTd concrete systems. The strategies include void filling, chemical re-alkalization, cathodic protection, chemical impregnation, and cable drying/de-humidification. Overall, this chapter is intended to provide researchers, engineers, and decision-makers with insights into corrosion mechanisms, inspection methods, and control strategies, ultimately aiming to enhance the corrosion-free service life of PTd concrete bridges.

Keywords Corrosion · Post-tensioned concrete · Bridge · Inspection · Repair · Service life

---

K. Manickam (✉)
Department of Civil Engineering, National Institute of Technology Rourkela, Rourkela, Odisha, India
e-mail: mkarthi@nitrkl.ac.in

C. Pichaimuthu
Department of Civil Engineering, National Institute of Technology Calicut, Calicut, Kerala, India
e-mail: chandru@nitc.ac.in

João M. P. Q. Delgado (ed.), *Advanced Materials, Corrosion Science, and Rehabilitation of Concrete Infrastructure*, Building Pathology and Rehabilitation 29,
https://doi.org/10.1007/978-3-032-28851-6_3

1 Introduction

The rapidly evolving urban infrastructure comprises a huge inventory of prestressed concrete, in which post-tensioned (PTd) concrete systems are widely used for constructing bridges, high-rise buildings, transmission pipes, etc. (Mohan et al. 2021). This chapter focuses only on the PTd concrete systems used for constructing bridges. PTd concrete systems offer more slender, aesthetically pleasing, and torsionally rigid structures (Haixue et al. 2022). The development of precast technology has furthered the use of PTd concrete systems in constructing highway, railway, and metro bridges. Typically, most of these bridges are designed for an anticipated corrosion-free service life of 100+ years (Mohan et al. 2021; Manickam and Pillai 2023a). However, there is a lack of stringent performance specifications for grouting materials and anchorage protection in many developing countries (Manickam and Pillai 2023a; Manickam and Pillai 2024a). Site-batched grout, still commonly used for grouting tendons, showed several issues (excessive bleeding, segregation, and soft grout formation) (Manickam and Pillai 2023a; Trejo et al. 2009). Furthermore, the lack of permanent anchorage protection systems (plastic caps, concrete pourbacks, waterproofing membrane, etc.) makes the anchorage zone susceptible to humidity and other deleterious elements (Manickam and Pillai 2023a). These problems can cause premature corrosion (within 20 years) of strands, which can result in the failure of tendons or even the entire structure (Trejo et al. 2009a).

Figure 1 shows a photo of an inadequately grouted anchorage from a bridge in a coastal city. The inset of the figure shows the incompletely filled anchorage exposed to the atmosphere. Likewise, Fig. 2 shows the photo of a corroded and broken tendon extracted from a 12-year-old bridge. It can be clearly seen that the top region of the tendon is severely corroded, whereas the bottom region remains uncorroded due to the presence of voids in the duct (Manickam and Pillai 2024b). Since the prestressing strands are enclosed within the duct, their condition cannot be observed externally. Consequently, visual inspections are inadequate for reliably assessing the corrosion in PTd concrete bridges. If undetected, corrosion in the anchorage zone may reduce the load-bearing capacity and cause catastrophic failures (Trejo et al. 2009b). Decommissioning of the structure will lead to significant wastage of steel and concrete, thereby directly impacting the carbon footprint and sustainability of the built environment (Iannuzzi and Frankel 2022; Chatterjee 2013). According to the NACE Impact Report 2016, the costs associated with the corrosion and repair of concrete structures account for approximately 4–5% of the global GDP (Koch et al. 2016). The primary challenge in the repair and maintenance of PTd concrete bridges lies in the difficulty of accessing the strands, which are embedded within the ducts inside the concrete. Inspection and repair operations must be conducted externally through the anchorage. Notably, the conventional repair method of void filling has caused the corrosion failure of a tendon in a PTd concrete bridge within four years after repair, due to the electrochemical incompatibility between the old and repair grouts (Trejo et al. 2009a). Hence, there is a need to find more robust and electrochemically viable repair solutions for controlling corrosion in PTd bridges.

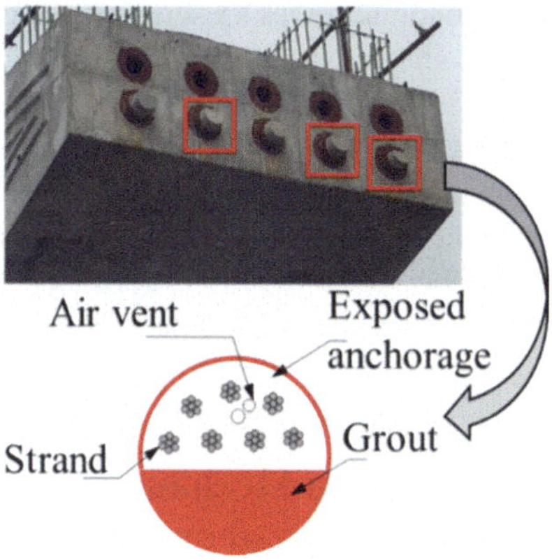

Fig. 1 Inadequately grouted anchorage from a bridge located in a coastal city

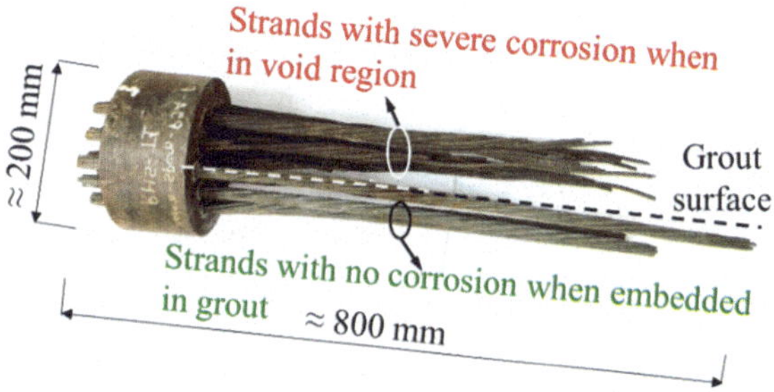

Fig. 2 Corroded and broken tendon from a 12-year-old bridge in a metro city (adapted from Manickam and Pillai 2024a)

This chapter provides a comprehensive review of the various corrosion mechanisms, inspection methods, and corrosion control strategies aimed at improving the service life of grouted PTd concrete bridges. The review focuses solely on the anchorage zone of bonded PTd concrete bridges, excluding unbonded and external tendon systems. The remainder of the chapter is organized as follows: First, the need for progressive research in this area is highlighted. Then, a review of the corrosion mechanisms in the anchorage region of PTd concrete bridges is provided. Then, a review of the site-inspection and condition assessment methods and their challenges is presented. Finally, a review of the various corrosion control and prevention strategies developed to improve the service life of PTd concrete systems and their limitations is provided.

2 Need for Progressive Research in this Area

Over the last few decades, many bridges have been built by developing countries because of the construction boom. These bridges are relatively young and may start showing signs of corrosion because of the inadequate quality grouting materials, construction practices, and anchorage protection systems. These bridges are important assets to a country, and failure or decommissioning could cause significant safety and economic losses. A series of accidents that happened worldwide in the last few decades have highlighted the severity of tendon corrosion and its consequences on PTd bridges (Trejo et al. 2009a). As the bridge infrastructure enters the era of maintenance and repair, it is vital for the engineering community to be ready with the technical expertise and practical skills necessary to assess and control corrosion in PTd concrete bridges. To achieve this goal, this chapter provides a comprehensive review of the various corrosion mechanisms, inspection methods, and repair strategies for the anchorage regions of grouted, PTd concrete systems.

3 Corrosion Mechanisms in Grouted, Post-tensioned Concrete Systems

Prestressed concrete (PC) structures can be categorized into two main types: (i) pretensioned concrete (PT) and (ii) post-tensioned (PTd) concrete. In PT concrete, the strands are tensioned first, followed by the casting of concrete. The stress transfer will happen through the bond between the strand and the concrete. Whereas, in a PTd concrete, the concrete is cast first, followed by the stressing of strands. PTd concrete can further be classified into two types: bonded systems and unbonded systems. In bonded systems, the ducts encasing the strands are filled with cementitious grouts, while in unbonded systems, the strands are coated with grease/wax without any filling of the encasing duct. In both systems, the stress transfer occurs mainly through the bearing action at the end anchorages. However, a small amount of stress transfer can happen through the bond between the strand and the grout for a bonded system. Furthermore, PTd concrete can also be divided into internal and external systems based on the location of the tendons. When the tendons are situated inside the concrete, it is referred to as an internal system, and vice versa (Lau and Lasa 2016). The prestressing strands are made of cold-drawn high-carbon steel and are more susceptible to corrosion than mild steel rebars because of the higher energy stored in the material during the manufacturing process (Haixue et al. 2022). Figure 3 shows the schematic of a PTd concrete bridge. The inset shows an inadequately grouted anchorage region exposed to atmospheric humidity and other deleterious elements that could cause corrosion. The possible mechanisms that can cause corrosion in the anchorage zone are presented below.

3.1 Corrosion Due to Inadequate Grouting Material

In PTd concrete systems, the tendons are filled with cementitious grout and sealed with a grout cap. This grout helps maintain a high alkaline environment around the tendons, providing protection against corrosion (Lau and Lasa 2016). Additionally,

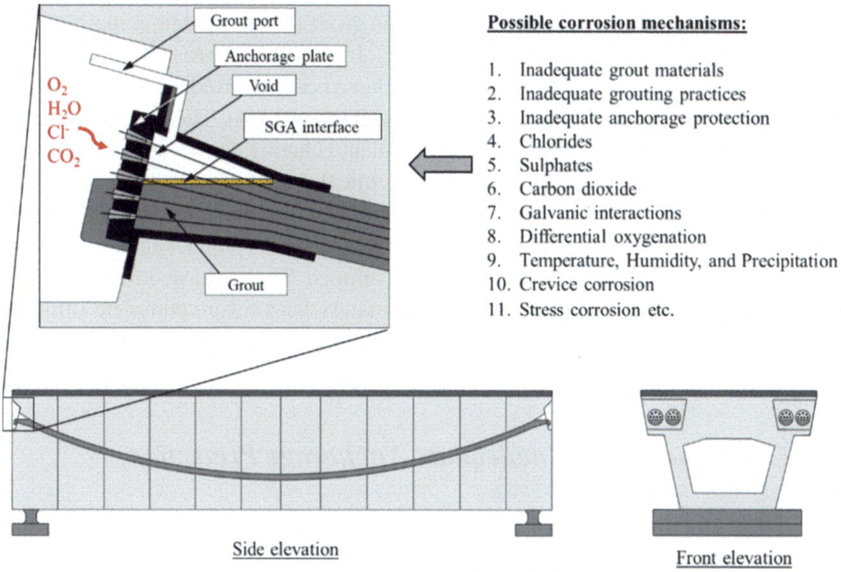

Fig. 3 Schematic of an inadequately grouted post-tensioned concrete bridge

the grout enhances stress transfer through surface bonding. However, the use of inadequate materials and inadequate workmanship in grouting has been identified as a cause of deficiencies, including (i) voids, (ii) segregated grout, and (iii) soft grout (Salas et al. 2004; Trejo et al. 2009c; Kamalakannan et al. 2018; Mohan et al. 2023). The Post-tensioning Institute (PTI) categorizes grouts into four distinct classes. Class A grouts consist of cement and water; Class B grouts comprise cement, water, and mineral/chemical admixtures; Class C grouts are pre-packaged, containing engineered grout materials and water; and Class D grouts are specialized formulations designed by the engineer on-site (PTI 2003). Prior to the early 2000s, Class A or Class B grouts were predominantly employed as the grouting material for most PTd bridges (Trejo et al. 2009a). These grouts typically exhibit a bleed of 3–5% by volume, with the bleed water potentially rising to the highest points of the ducts (Lau et al. 2013a, 2013b). Additionally, the capillary rise of water through the narrow interstitial spaces in the strands can exacerbate bleeding (Powers et al. 2016). Once the bleed water evaporates or is absorbed into the underlying grout, voids can form near the anchorage regions. These voids, due to the openings in the anchorage regions, can expose the prestressing steel to oxygen, moisture, chloride, and carbon dioxide, thereby initiating and sustaining corrosion in the void regions (Manickam and Pillai 2023a). With the advent of pre-packaged thixotropic grouts like Class C and Class D, voids caused by grout bleed have been significantly minimized (Trejo et al. 2009a). However, these grouts could also bleed and segregate (Sprinkel 2015). Such segregation can result in grout deficiencies, including elevated moisture content, pH, and sulfate levels (Vigneshwaran et al. 2018). Additionally, pockets of soft grout may form. Soft grout refers to a condition where the grout fails to harden, remaining soft and permeable after setting (Mohan et al. 2021). Instances of corrosion failure in PTd tendons using pre-packaged thixotropic grout have been reported in Florida bridges, even within eight years of service (Sagues and Powers 2006). These issues in grout materials can lead to variations in their physical and chemical properties. Changes in properties such as pH, density, porosity, and chemical composition can result in differences in corrosion potentials, initiate macrocell formation, and cause corrosion (Lau et al. 2013a, 2013b). To control this issue, high-performance prepackaged thixotropic grouts with zero bleed resistance have been developed. Studies have developed stringent performance specifications for grouting materials to ensure complete filling of the ducts and anchorages (Manickam and Pillai 2023a; Mohan et al. 2023).

3.2 Corrosion Due to Inadequate Anchorage Protection

Improper and inadequate anchorage protection systems have been used in many bridge projects, leading to several problems. In many cases, insufficient grouting and sealing of the anchorage have been observed, often using only a cementitious cap without any plastic or metallic cover, as shown in Fig. 1. These cementitious materials are hygroscopic, and they can attract and retain humidity, which promotes corrosion of the metallic components within the anchorages (Manickam and Pillai 2023a). In some instances, polymeric coatings have been used as waterproofing over the grout cap without any concrete cover. Unfortunately, these coatings are vulnerable to UV exposure and can degrade quickly, leaving the anchorage zone exposed to humidity (Ching et al. 2018). Furthermore, the faulty design of the rainwater drain situated directly beneath the construction joint can also lead to severe corrosion issues (FDOT 2001). Manickam and Pillai have presented case studies on inadequate grouting practices and anchorage protection and highlighted the need for stringent specifications (Manickam and Pillai 2023a).

3.3 Corrosion Due to Chlorides in the Grout

Chlorides can infiltrate the duct from the atmosphere or surface runoff water, subsequently accumulating in void regions. Additionally, minor quantities of chlorides (> 50 ppm) may be present in the original grout composition (Sagues and Powers 2006). ACI reports that the permissible level of acid-soluble chlorides in grouts should not exceed 0.08% by weight of cement (ACI 222R-01 2010). Due to the evaporation of bleed water, chlorides tend to accumulate on hardened grout surfaces, potentially exceeding the threshold necessary to initiate corrosion (Taveira et al. 2008). Research indicates that corrosion may start when chloride levels exceed a threshold range of 0.4–0.8% by weight of cement (Lee 2014), although there is considerable variability in these reported values. Once chlorides penetrate the duct, the minimal grout cover is insufficient to prevent chlorides from reaching the steel, regardless of the resistance of the grout (Yong-Sik et al. 2018). In addition, chlorides facilitate the binding of oxygen atoms to iron atoms on the steel surface, thereby exacerbating corrosion reactions. At the grout surface, elevated chloride content and reduced pH levels (due to carbonation) significantly increase the $[Cl^-]/[OH^-]$ ratio compared to the bulk of the grout (Sagues and Powers 2006). This condition is adequate to disrupt passivity and initiate pitting on the steel surface (Taveira et al. 2008). Exposure to chlorides for one year can reduce the tensile capacity of PTd strands by up to 27% (Manickam and Pillai 2024b).

3.4 Corrosion Due to Sulphates in the Grout

In addition to chloride content, segregation increases the local sulfate content in the grout. An excess of sulfate can reduce the binding of chloride ions by the grout (Permeh et al. 2019). A study indicated that the sulfate content in grout should be restricted to 0.07% by the weight of the grout powder (Permeh et al. 2019; Permeh et al. 2018). Consequently, corrosion can be influenced by the combined effects of chloride and sulfate ions. Beyond the strands, chlorides can hasten the corrosion of various metallic components at the anchorages. Notably, the bearing plate is more prone to chloride-induced corrosion than the wedges (Li et al. 2017).

3.5 Corrosion Due to Carbonation of Grout

Typically, cementitious composites comprise 25–50% calcium hydroxide and have a pH exceeding 13, which is adequate to establish an alkaline environment conducive to the passivation of embedded strands (Fuhaid and Niaz 2022). Upon carbonation, the pH of the grout decreases to 9 or lower, with fully carbonated grout potentially reaching a pH as low as 7 (Manickam and Pillai 2024b). In the PT duct of the Mid Bay Bridge in Florida, the grout was carbonated to a depth of approximately 6 mm (Salas et al. 2004). The Strand-Grout-Air (SGA) interface is the region where the strand transitions from the hardened grout into the open-air voids, as shown in Fig. 3. Depassivation of strands may occur at the SGA interface due to grout carbonation (Wang et al. 2005). A study examined the impact of various parameters, including grout class, moisture content, chloride concentration, void type, and stress level, on strand corrosion and found that corrosion was most severe at the SGA interface (Trejo et al. 2009a). The SGA interface exhibits a higher $[Cl^-]/[OH^-]$ ratio compared to the bulk of the grout, sufficient to depassivate the steel (Powers et al. 2016). Once corrosion initiates at the SGA interface, a galvanic coupling occurs between the relatively small active area (SGA interface) and the larger passive area of the strand embedded in the grout (Powers et al. 2016). In addition to carbonation, the hydrolysis of metal ions further reduces the pH of the grout (Powers et al. 2016).

3.6 Corrosion Due to Galvanic Coupling of Metals

Typically, the anchorage of a PTd system comprises a ductile cast iron anchor head, forged steel wedge plate, wedges, and high-strength steel strands covered with a cementitious cap (Theryo et al. 2013). Coupling between these components could significantly aggravate galvanic corrosion. The cementitious grout has sufficient conductivity to support the galvanic action, and external water intrusions can accelerate the corrosion process (Wang et al. 2005). Studies reported that the strands are mostly anodic to the other metallic components in the anchorage (Powers et al. 2016; Sagues and Powers 2006; Wang et al. 2005; Sagues et al. 2003). Laboratory experiments reported that the anodic charge supplied by the strand was sufficiently high to develop first wire failures within two to three years of exposure (Sagues and Powers 2006). In addition to the strands, the anchor head and bearing plates are also vulnerable to galvanic corrosion. Dickson et al. reported the severe corrosion of the anchor head and bearing plate of a 34-year-old PTd girder (Dickson et al. 1993). Anchorage corrosion may lead to the entry of moisture into the ducts, facilitating the corrosion of the strands inside the duct. During the process of post-tensioning, the prestressing strand is locked into the wedges at both ends of a PTd element to carry the prestress. During load transfer, there will be indentations or shear markings made on the strand surface due to the sharp wedge teeth. These indentations act as small crevices and initiate wedge-zone corrosion (Hartt et al. 2015). The corrosion is more severe near the tail part of the wedge compared to the head part. The reason is that the tail part of the wedge is located near the exposed anchorage and has a higher oxygen supply compared to the head part (Li et al. 2017). In some cases, the wedge teeth may also corrode. The wedge-zone corrosion may lead to disengagement, slippage, load release, and tendon failure, affecting the integrity of the structure (Hartt et al. 2015). Spiral reinforcements are provided near the anchorage zone to carry the incoming bearing stresses. There could be limited galvanic corrosion between the spiral reinforcements and the strands and between the spiral reinforcements and the bearing plate, respectively (Pillai 2009). Another study reported severe galvanic corrosion of carbon steel ducts when coupled with the prestressing strands (FLDOT 2002).

3.7 Corrosion in Re-filled Ducts

The case study concerning tendon failure in a bridge located in Virginia illustrates the phenomenon of galvanic corrosion-induced tendon failure. A PTd bridge in Virginia exhibited voids and strand corrosion. As a remedial measure, these voids were filled with high-performance repair grouts. Approximately four years subsequent to the repair, the tendon failed at the interface between the existing grout and the newly injected repair grout (Hanson 2007). The distinct physical and chemical properties of the two grouts resulted in the formation of a macro couple. Consequently, galvanic cells were established between the existing grout and the repair grout, leading to a type of galvanic corrosion, wherein a metal exposed to two dissimilar environments undergoes corrosion (Trejo et al. 2009a). Lau et al. conducted laboratory experiments to investigate the corrosion mechanisms in PTd tendons with voids repaired using dissimilar grout (Lau et al. 2012; Lau et al. 2013a, 2013b). The results indicated no evidence of severe corrosion. In contrast, Manickam and Pillai performed experimental and simulation studies and reported the possibility of significant galvanic corrosion and strand failures within 10 years (Manickam and Pillai 2023b). However, due to the limited information available, no definitive conclusions can be drawn, highlighting the need for further research to better understand this corrosion mechanism.

3.8 Corrosion Due to Differential Oxygenation

One significant factor influencing corrosion is the presence of oxygen, which determines the rate of the cathodic reaction. Oxygen is more readily available near the anchorage zone compared to the main body of the duct. This differential aeration process can result in localized corrosion (Lau et al. 2013a, 2013b). The top three to four strands nearest to the grout port are predominantly affected by corrosion (Reis 2007). Similarly, the interstitial space present in the helically wound prestressing strand creates small crevices, where the availability of oxygen is low compared to the outer surface of the strand. This can also aggravate crevice corrosion (Lau et al. 2013a, 2013b). Sagues et al. (2003) developed a mathematical model and reported that oxygen availability is the main factor that governs the corrosion process of strands in a PTd duct (Sagues et al. 2003).

3.9 Other Factors

Temperature can influence the rate of corrosion in steel, as electrochemical reactions occur more rapidly with rising temperatures (Konovalova 2021). External relative humidity directly impacts corrosion rates, which tend to increase exponentially beyond 60 °C (Haynic and Upham 1972). Additionally, the duration of precipitation affects the wet-dry cycles within the duct, potentially leading to corrosion (Trejo et al. 2009a). However, a comprehensive report on this topic is lacking in the literature. In summary, corrosion is influenced by the combined effect of external relative humidity, temperature, and precipitation during both the initiation and propagation phases. Moreover, the high stress in tendons, together with these factors, can result in stress-induced corrosion cracking (Kovac et al. 2007). The techniques used to identify and assess corrosion in the anchorage regions are described below.

4 Inspection and Condition Assessment Methods

Visual inspection may not effectively detect the condition of strands because these anchorages are embedded within concrete pourbacks (Moore et al. 2001). Assessment techniques can be categorized as non-destructive (non-invasive) and semi-destructive (invasive). They often involve the detection of strands, grout defects, or corrosion conditions. A common method of semi-destructive testing involves breaking pinpoints in the ducts to assess the condition of the strands and check for the presence of voids (Rapaport 2022). Samples from the assessment can be analyzed to determine the pH, chloride, and sulfate contents, which help identify the root causes of corrosion (Haixue et al. 2022). Half-cell potential measurements and corrosion rate measurements can be taken through these pinpoint locations to assess the corrosion condition of strands (Office of Bridges and Structures 2018, FHWA-HIF-18-029). Borescopes can be utilized to assess the condition of grout and strands inside the duct. Additionally, the screwdriver penetration test can evaluate the integrity of the strands and the level of prestress (MacDougall and Li 2007; Brenkus et al. 2021). However, quantitative information may not be obtained from this method. The radiological techniques using X-rays/Gamma rays can effectively detect the condition of voids, strands, and ducts. However, this method requires expensive equipment and skilled labor, making it unsuitable for all field conditions. Furthermore, the radiation involved poses safety concerns (Brenkus et al. 2021; Bridge-FHWA InfoTechnology 2026b, FHWA-RAD). The magnetic flux method is gaining widespread importance for assessing the corrosion condition of strands. This technique involves inducing a magnetic field near the tendons and analyzing their condition (Quaglini et al. 2023; Bridge-FHWA InfoTechnology 2026a, FHWA-MFL). However, it requires placing the sensor within 6 inches of the tendons to obtain a strong signal (Brenkus et al. 2021). Additionally, the presence of mild steel reinforcements and metallic ducts can reduce the accuracy of this method (Brenkus et al. 2021). Infrared thermography involves assessing the thermal properties of the grout to locate the presence of voids and the properties of the grout. The heat source can be either natural or applied (using infrared guns) (Azizinamini and Gull 2021). This technique is relatively simple and has been successfully applied for evaluating the condition of reinforced concrete, but it is quite challenging for assessing the condition of PTd anchorage, especially when the cover depth is large (FHWA-HIF-18-0292018). In the Ground Penetrating Radar (GPR) method, electromagnetic waves are transmitted at high velocity to detect features such as tendon profiles, duct conditions, voids, and the state of strands (Rossi et al. 2024). This method can accurately provide data up to a depth of 24 inches (Brenkus et al. 2021). However, the presence of congested reinforcement can reduce the accuracy of the results (FHWA-HIF-18-0292018). The Ultrasonic Pulse Echo method involves sending ultrasonic pulses through concrete to assess the condition of the grout and strands within the duct (Quaglini et al. 2023). However, it does not provide information about corrosion. This method can detect the presence of voids and the condition of the strands up to a depth of 12 inches (Brenkus et al. 2021). The Impact Echo method, on the other hand, involves sending acoustic waves through the material created by the impact of a steel ball, with the resulting reflected waves being measured (Quaglini et al. 2023). This technique can become tedious when assessing large areas without automation, especially in locations with dense reinforcements (Brenkus et al. 2021). Also, this method could fail to work in PTd systems with high-density polyethylene (HDPE) ducts (Iyer et al. 2002). In addition to the methods mentioned above, there are several other techniques available, such as the Acoustic Emission method, Time Domain Reflectometry, Magnetostrictive sensing, Electrochemical Methods, and so on (Brenkus et al. 2021). The effectiveness of the methods mentioned above is largely influenced by the local site conditions, the availability of advanced technologies, skilled labor, and budget constraints. These techniques are typically used for non-routine inspections only when a problem arises. However, it is essential to conduct regular and periodic condition assessments to prevent the initiation of corrosion and extend the service life of structures. After identifying the grout defect or corrosion, repairs must be made. The available repair techniques and their challenges are presented below.

5 Repair Techniques and Their Challenges

The maintenance and repair of the anchorage region includes addressing one or more of the following components: (i) prestressed strands, (ii) anchor head (bearing plate, wedges, and metallic trumpet), (iii) metallic/HDPE duct, (iv) grout cap, (v) deficient grout, (vi) concrete pourback, (vii) waterproofing coating, and (viii) mild steel reinforcements. The choice of selecting a repair method depends on several factors, including the type of PTd system, the type and location of the distress, the accessibility of the affected area, and the availability of specialized personnel and equipment (Sharma et al. 2024). It is essential to carefully select the appropriate repair method, as improper choices can lead to accelerated corrosion and subsequent failures. Upon inspection and condition assessment, the following strategies can be considered: (i) repair, (ii) partial or complete replacement, (iii) strengthening, and (iv) decommissioning (Jimenez 2013). In cases where corrosion is localized, a partial replacement of tendons may be appropriate. However, if multiple strand failures occur, full replacement of the structure will be necessary since internal tendons cannot be replaced. For severe damage, external post-tensioning can be utilized to strengthen the structure (Brenkus et al. 2021). This review focuses solely on repair methods for corrosion and excludes structural strengthening techniques. Traditional repair methods typically involve replacing the deficient grouts (segregated and soft grouts) in the duct with new ones. However, concerns have been raised about potential corrosion in PTd bridges following such repairs (Manickam and Pillai 2024b). Consequently, additional methods have been developed to control corrosion in PTd concrete systems. Below is a summary of some relevant techniques for repairing PTd anchorage regions and the challenges associated with them.

5.1 Void Filling

Filling voids with cementitious grout is among the most practical methods for repairing a PTd anchorage system. While extensive research has been conducted on the initial grouting of ducts, studies on repair-grouting remain limited. Void filling can be achieved through i) pressure grouting, ii) vacuum grouting, and iii) pressure-vacuum grouting. Class C grouts, which are pre-packaged, can serve as repair grouts (Im et al. 2010). Filling capability refers to the effectiveness of the selected repair method-material combination in filling voids (Im et al. 2010). Im et al. evaluated grouting methods based on infiltration length measurements and recommended the pressure-vacuum grouting method as particularly effective for void filling (Im et al. 2010). Another study recommended the use of tremie grouting for smaller voids (about 1 m length) and vacuum grouting for larger voids (more than 1 m) (Ganz and Vildaer 2002). However, it is crucial to ensure the electrochemical compatibility between the repair grout and the parent grout before application (Manickam and Pillai 2024a). Manickam and Pillai recommended re-alkalizing the carbonated SGA interface before regrouting to control the galvanic corrosion (Manickam and Pillai 2024b). Carbonated grout can be re-alkalized through capillary absorption of highly alkaline solution, electrochemical re-alkalization (ERA) techniques, or by using corrosion inhibitors. Re-alkalization eliminates the need to replace carbonated areas and can elevate the alkalinity to 12 or higher (Goyal et al. 2018). A study recommended the use of a 1 M $Ca(OH)_2$ solution for seven days to effectively re-alkalize the carbonated grout and re-passivate the corroded strands (Manickam and Pillai 2024b). The rate at which the alkaline solution penetrates is influenced by the moisture content of the pores; the drier the condition, the greater the absorption (Abdelaziz and Page 2004). The process of re-alkalization through the absorption and diffusion of the alkaline solution has been relatively underexplored in scientific literature. Another method is the ERA of carbonated grouts. Nonetheless, the method is complicated, and there are potential risks of hydrogen-assisted environmental cracking and bond loss resulting from the passage of high currents through the strands (NACE-01104 2020). A study proposed the injection of rust inhibitor aqueous solution (such as lithium nitrite) into the void region before grouting. This process helps to passivate the metal. The study also recommends the mixing of grout material with a rust preventive agent (such as lithium nitrate or calcium nitrate) to prevent corrosion (Kobe University 2011). However, completely filling all voids in the duct is a challenging task (Miller et al. 2017).

5.2 Cable Drying/De-Humidification

The cable drying/de-humidification method is mainly used for unbonded suspension bridges (Bloomstine et al. 2006). However, cable drying can also be used for repairing soft grout in bonded PTd systems. In this method, dry air is passed through the gaps in the tendons in the longitudinal direction to remove the moisture content in the strands and the embedded grout. Studies reported that drying could reduce the moisture content in the softgrout by less than 1% (Patil et al. 2019; Patil et al. 2020). However, this method could not be effective if there are cracks in the ducts and the grout is impervious (Lee 2022). Furthermore, drying could result in shrinkage and cracks in the grout. In addition, macrocell corrosion was also reported (Patil et al. 2019; Patil et al. 2020). This may be due to the porosity of the dried grout attracting oxygen and moisture from the air. This method has been revised by injecting inert gas (Nitrogen), which could reduce the chance of corrosion (Lee 2022). A study used high-pressure water jets to remove soft and segregated grout in ducts. However, the hydrodemolition technique was unable to completely eliminate all defective grout (Patil et al. 2020).

5.3 Chemical Impregnation

In this method, a corrosion-inhibiting chemical is injected through the interstitial spaces in the strands along the longitudinal direction. This chemical, through capillary action, flows through the gaps and forms a coating over the steel, while also enhancing the resistance of the surrounding grout. A study reported that chemical impregnation can reduce corrosion by over 90% (Haixue et al. 2022). Another study assessed the corrosion resistance of the impregnated material by potentiostatic and galvanostatic testing and reported the significant improvement after impregnation. The study also reported the successful field application of this method (Whitmore and Lasa 2018). However, the presence of water in the ducts could emulsify the chemicals and reduce their efficiency (Miller et al. 2017).

5.4 Cathodic Protection

Cathodic protection (CP) is an electrochemical repair technique that involves polarizing the metal to be protected towards the cathodic direction. CP of a metal can be achieved by (i) impressing an external current and (ii) connecting to another sacrificial metal (say, zinc, aluminum, or magnesium) (Pedeferri 1996; Angst 2019). The concept of CP for prestressed concrete bridges has been derived from its use in prestressed concrete cylinder pipes for water and sewer transmission services (Hall 1998). CP is typically employed in bridges to safeguard the mild steel reinforcements against corrosion (Hondel et al. 2018). However, the application of CP to the prestressing strand remains in a developmental phase due to several concerns, which are discussed below. Prestressed steel is susceptible to hydrogen embrittlement (HE) when they are cathodically overprotected (El-Amoush et al. 2018). The reason could be the unavailability of oxygen at the cathode surface due to the slower rate of oxygen diffusion through the highly resistive cover concrete, which thermodynamically favors the hydrogen evolution reaction. In addition, several investigations reported that there could be bond loss due to (i) long-term cathodic polarization and (ii) over-protection of the prestressing steel. The reason is the deterioration or weakening, or softening of the cement phases around the steel interface due to the cathodic polarization-induced electromigration (Chang et al. 1999; Ducrocq 1990). The governing parameters for the bond loss are (i) the charge transfer density and (ii) the total charge transferred. The reduction in the bond can compromise structural integrity and reduce the anticipated service life of the structure (Ducrocq 1990). However, the magnitude of bond loss is not clearly defined in the literature. Other concerns associated with CP of PTd systems are (i) barrier effect due to the metallic/plastic ducts, and (ii) stray current corrosion when proper electrical continuity is not achieved. However, CP using a galvanic anode is considered safe from the standpoint of HE due to the limited polarization and ohmic losses in concrete structures (NACE-01102 2002). Currently, there are no established procedures for designing CP systems for PTd concrete. The criteria outlined in the existing standard ISO 12696 are not robust enough to be applied to the design of CP for PTd anchorages (ISO-12696 2022). Limited work has been reported on the CP of PTd strands. A finite element model was used to understand the feasibility of impressed current cathodic protection grouted PT systems as a preventive repair strategy. The model assumed a line anode running parallel to the tendons inside the duct, and predicted sufficient polarization of the strands without any grout acidification (Bumgardner 2015). A study reported that arc-sprayed galvanic anodes could not develop enough current densities to arrest corrosion in a highly resistive PTd system (Gawedzinski 2002). However, NACE 01102 recommends that galvanic anodes can be used to control the corrosion of anchors and strands in PTd systems (NACE-01102 2002) Some studies have implemented the cathodic protection of the anchorage regions using activated arc-sprayed zinc anodes. Also, galvanic anodes have been used to protect the tendons inside the duct (Haixue et al. 2022). However, these studies lack a scientific rationale for design. Manickam and Pillai proposed the CP of the anchorage regions using discrete galvanic anodes. They established that galvanic anodes could protect the inside and outside regions of the anchorage systems (Manickam and Pillai 2024b). In addition to these techniques, understanding corrosion prevention methods is essential, and they are discussed below.

6 Corrosion Prevention Strategies

It is very important to prevent corrosion rather than to control it. The use of proper grouting materials and adequate anchorage protection systems shall prevent the initiation of corrosion. High-performance prepackaged thixotropic grouts with zero bleed resistance, high fluidity, high carbonation resistance, and low shrinkage shall be used to grout the tendons (Manickam and Pillai 2023a). The PTd anchorage region shall be protected with multi-level corrosion protection systems. This way, if one system fails,

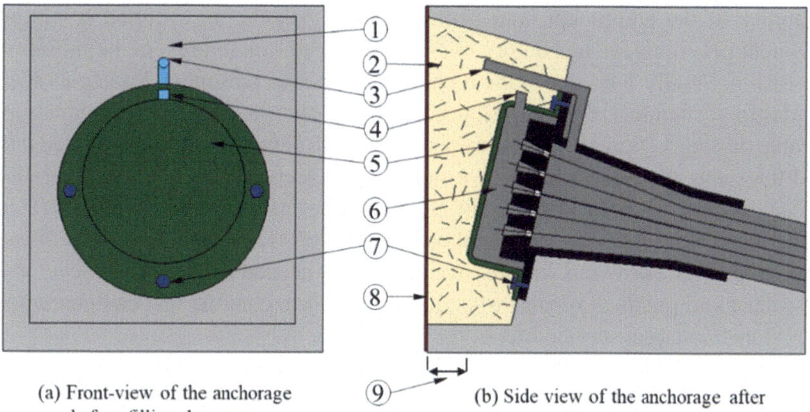

(a) Front-view of the anchorage before filling the recess

(b) Side view of the anchorage after filling the recess

Note:
1 – Clearance for tendon grout port; 2 – Fibre-reinforced cover concrete;
3 – Tendon grout port (diameter ≥ 30 mm); 4 – Air vent; 5 – Transparent acrylic cap;
6 – Grout filling; 7 – Stainless steel screw; 8 – Waterproofing layer; 9 – Cover for acrylic cap ≥ 50 mm

Fig. 4 Multi-level anchorage protection systems (adapted from Manickam and Pillai 2024b)

the other will still safeguard the strands. Ideally, the exposed anchorage should be covered with plastic or metallic caps, sealed with a waterproofing membrane, and then enclosed with concrete pourbacks. Additionally, the deck slab should be designed to prevent water from dripping into the anchorages (Manickam and Pillai 2024a). Figure 4 shows the multi-level corrosion protection systems proposed by Manickam and Pillai. Li et al. provided detailed guidelines for the selection of multi-layer corrosion protection systems for the anchorage regions exposed to various environmental conditions (Li et al. 2008). Also, PTI has proposed performance specifications for effective grouting practices (PTI 203). Apart from the conventional strands, stainless steel and hot-dipped galvanized strands could offer higher corrosion resistance (Li et al. 2008).

7 Summary and Conclusions

This paper presents a comprehensive review of the various corrosion mechanisms, inspection methods, and repair strategies for the anchorage regions of grouted, post-tensioned (PTd) concrete bridges. The mechanisms include corrosion due to chlorides, sulphates, carbon dioxide, oxygen, galvanic interactions, grout deficiencies, and inadequate anchorage protection systems. Then, a review of the various inspection and condition assessment methods was discussed. The review highlights the challenges associated with each of the methods. It could be concluded that the choice of selecting an inspection method depends on the site conditions, budget, availability of the equipment, and skilled labour. Periodic and routine condition assessment programs are mandatory to assess the condition to prevent the initiation of corrosion. Finally, a review of the repair strategies was presented. Strategies such as void filling, chemical re-alkalization, cathodic protection, chemical impregnation, and cable drying/de-humidification were discussed in detail. It can be concluded that void filling was the easiest method; however, the electrochemical incompatibility between the parent grout and the repair grout must be ensured to prevent accelerated galvanic corrosion. Also, galvanic cathodic protection is a non-invasive, promising method to prevent corrosion. Progressive research is needed in this area to develop a more robust and practical repair inspection and repair technique that can guarantee a corrosion-free target service life.

References

Abdelaziz GE, Page CL (2004) Factors controlling the mechanism of electrochemical realkalisation. Eng Res J 1:16–30

ACI 222R-01 (2010) Protection of metals in concrete against corrosion. American Concrete Institute, Farmington Hills, USA

Angst UM (2019) A critical review of the science and engineering of cathodic protection of steel in soil and concrete. Corrosion 75:1420–1433. https://doi.org/10.5006/3355

Azizinamini A, Gull J (2021) Improved inspection techniques for steel prestressing/post-tensioned strand, vol 1. FLDOT, Miami, USA

Bloomstine ML, Sorensen O, Thomsen JV (2006) Main cable corrosion protection by dehumidification. In: Paper presented at the IABSE conference, Copenhagen. https://doi.org/10.2749/222137806796236231

Brenkus N, Tatum G, Kreitzer I (2021) Repair and maintenance of post-tensioned concrete bridges. In: National Cooperative Highway Research Program (NCHRP) synthesis 562. National Academy of Sciences, Washington, DC. https://doi.org/10.17226/26172

Bridge-FHWA InfoTechnology (2026a) Bridge-Magnetic Flux Leakage (MFL). https://infotechnology.fhwa.dot.gov/magnetic-flux-leakage-mfl/. Accessed on July 13 2026

Bridge-FHWA InfoTechnology (2026b) Bridge-Radiography (RAD Tendons). https://infotechnology.fhwa.dot.gov/radiography-rad-tendons/. Accessed on July 13 2026

Bumgardner JD (2015) Feasibility of cathodic protection in grouted post-tensioned tendons-exploratory model calculations. Dissertation, University of South Florida

Chang JJ, Yeih W, Huang R (1999) Degradation of the bond strength between rebar and concrete due to the impressed cathodic current. J Mar Sci Technol 7:89–93

Chatterjee AK (2013) Concrete repair materials: polymers and green chemistry—how far synergistic are they? Int J 3R's 4:534–538

Ching YC, Udenni Gunathilake TMS, Ching KY, Chuah CH, Sandu V, Singh R, Liou NS (2018) Effects of high temperature and ultraviolet radiation on polymer composites. In: Durability and life prediction in biocomposites, fibre-reinforced composites and hybrid composites. https://doi.org/10.1016/B978-0-08-102290-0.00018-0

Dickson TJ, Tabatabai H, Whiting DA (1993) Corrosion assessment of a 34-year-old precast post-tensioned concrete girder. PCI J 38:44–51

Ducrocq L (1990) Effect of cathodic protection on bond strength between prestressing steel and concrete. Dissertation, Florida Atlantic University

El-Amoush AS, Al-Duheisat SA (2018) Cathodic polarization behavior of the structural steel wires under different prestressing conditions. J Mater Res Technol 7:1–6. https://doi.org/10.1016/j.jmrt.2016.10.004
FDOT (2001) Mid-Bay bridge post-tensioning evaluation—Final Report. Corven Engineering, Inc., Florida Department of Transportation, Tallahassee, FL, USA
FLDOT (2002) Galvanic testing of stressed strands inside carbon steel pipe. Florida Department of Transportation, Miami, USA
Fuhaid AFA, Niaz A (2022) Carbonation and corrosion problems in reinforced concrete structures. Buildings 12:1–20. https://doi.org/10.3390/buildings12050586
Ganz HR, Vildaer S (2002) Grouting of post-tensioning tendons. VSL International Ltd., Switzerland
Gawedzinski M (2002) Evaluation of sprayed-on metalizing for precast prestressed concrete I-Beams. Report No. 141, IRBC Program, Illinois Department of Transportation, USA
Goyal A, Pouya HS, Ganjian E, Claisse P (2018) A review of corrosion and protection of steel in concrete. Arab J Sci Eng 43:5035–5055. https://doi.org/10.1007/s13369-018-3303-2
Haixue L, Jovan V, Whitmore D (2022) Post-tensioning investigation and corrosion mitigation in voided slab bridges. In: Paper presented at the 11th international conference on short and medium span bridges, Ontario
Hall SC (1998) Cathodic protection criteria for prestressed concrete pipe—an update. In: Paper presented at the CORROSION 98, San Diego, California. Paper No. 98637, NACE International
Hartt WH, Vincent FF, Ivanov VI (2015) Wire/strand slippage in anchorages—a potential failure mode for corroding post-tensioning tendons. In: Paper presented at the CORROSION 2015, Dallas, Texas. Paper No. 5461, NACE International
Haynie FH, Upham JB (1972) Correlation between corrosion behavior of steel and atmospheric pollution data. Corrosion Natural Environ. https://doi.org/10.1520/STP32155S
Hondel AJ, Gulikers J, Giorgini R, van den Hondel A(WM) (2018) A 5 year track record on a galvanic CP system applied on a light weight concrete bridge with prestressed steel—developments in time of the effectiveness as determined by depolarisation values and current densities. In: MATEC web of conferences, vol 199, pp 1–8. https://doi.org/10.1051/matecconf/201819905004
Iannuzzi M, Frankel GS (2022) The carbon footprint of steel corrosion. npj Mater Degrad 6:1–4. https://doi.org/10.1038/s41529-022-00318-1
Im SB, Hurlebaus S, Trejo D (2010) Effective repair grouting methods and materials for filling voids in external posttensioned tendons. Transp Res Rec 2172:3–10
ISO 12696 (2022) Cathodic protection of steel in concrete. ISO, Geneva, Switzerland
Iyer S, Schokker AJ, Sinha SK (2002) Ultrasonic imaging–a novel way to investigate corrosion status in post-tensioned concrete members. J Indian Inst Sci 82:197–217
Jimenez GA (2013) Guide for evaluation and repair of unbonded post-tensioned concrete structures. Concrete Repair Bulletin:16–19
Kamalakannan S, Thirunavukkarasu R, Pillai RG, Santhanam M (2018) Factors affecting the performance characteristics of cementitious grouts for post-tensioning applications. Constr Build Mater 180:681–691. https://doi.org/10.1016/j.conbuildmat.2018.05.236
Kobe University NUC (2011) Method for inhibiting corrosion of PC tendons. Patent No. JP5312526B2, Japan
Koch G, Varney J, Thompson NO, Moghissi O, Gould M, Payer J (2016) NACE international impact report. NACE International
Konovalova V (2021) The effect of temperature on the corrosion rate of iron-carbon alloys. Mater Today Proc 38:1326–1329. https://doi.org/10.1016/j.matpr.2020.08.094
Kovač J, Leban M, Legat A (2007) Detection of SCC on prestressing steel wire by the simultaneous use of electrochemical noise and acoustic emission measurements. Electrochemica Acta 52:7607–7616. https://doi.org/10.1016/j.electacta.2006.12.085
Lau K, Lasa I (2016) Corrosion of prestress and post-tension reinforced-concrete bridges. Corrosion Steel Concrete Struct:37–57. https://doi.org/10.1016/B978-1-78242-381-2.00003-1

Lau K, Paredes M, Rafols J (2012) Corrosion evaluation of post-tensioned tendons with dissimilar grout. In: Paper presented at the CORROSION 2012, Salt Lake City, Utah. Paper No. 1738, NACE International

Lau K, Lasa I, Paredes M (2013a) Corrosion failure of post-tensioned tendons in presence of deficient grout. In: Paper presented at the CORROSION 2013, Orlando, Florida. Paper No. 2600, NACE International

Lau K, Rafols J, Lasa I, Paredes M (2013b) Laboratory corrosion assessment of post-tensioned tendons repaired with dissimilar grout. In: Paper presented at the CORROSION 2013, Orlando, Florida. Paper No. 2602, NACE International

Lee SK (2022) Corrosion-induced durability issues and maintenance strategies for post-tensioned concrete bridges. Report No. FHWA-HRT-22-090. USDOT FHWA, USA

Lee SK, Zielske J (2014) An FHWA special study: post-tensioning tendon grout chloride thresholds. Report No. FHWA-HRT-14-039. USDOT FHWA

Li F, Luo X, Liu Z (2017) Corrosion of anchorage head system of post-tensioned prestressed concrete structures under chloride environment. Struct Concr 18:902–913. https://doi.org/10.1002/suco.201600140

Li K, Chen Z, Lian H (2008) Concepts and requirements of durability design for concrete structures: an extensive review of CCES01. Mater Struct 41:717–731. https://doi.org/10.1617/s11527-007-9276-3

MacDougall C, Li S (2007) Determining broken wires in unbonded seven-wire strands using penetration tests. PCI J 5:96–104. https://doi.org/10.15554/pcij.09012007.96.104

Manickam K, Pillai RG (2023a) Galvanic corrosion and cathodic protection of re-grouted post-tensioned concrete systems. In: MATEC web of conferences, vol 378. https://doi.org/10.1051/matecconf/202337807002

Manickam K, Pillai RG (2023b) Grouting materials and practices for century-long corrosion protection of post-tensioned concrete bridges. Indian Concrete J 97:6–20

Manickam K, Pillai RG (2024a) Galvanic corrosion of strands in re-grouted, post-tensioned concrete bridges. Corrosion 80:130–141. https://doi.org/10.5006/4461

Manickam K, Pillai RG (2024b) Chemical and electrochemical repair methods for the anchorage regions of grouted, post-tensioned concrete systems. Structure 69:107595. https://doi.org/10.1016/j.istruc.2024.107595

Miller EL, White BC, Haskins RW, Ebeling RM, Evans JA (2017) An investigation of corrosion mitigation strategies for aging post-tensioned cables Information Technology Laboratory. In: Report no. ERDC/ITL TR-17-1. The US Army Engineer Research and Development, Center, USA

Mohan MK, Pillai RG, Santhanam M, Gettu R (2021) High-performance cementitious grout with fly ash for corrosion protection of post-tensioned concrete structures. Constr Build Mater 281:122612. https://doi.org/10.1016/j.conbuildmat.2021.122612

Mohan MK, Manohar S, Pillai RG, Santhanam M, Gettu R (2023) High-performance cementitious grouts for post-tensioned concrete systems–performance specifications and prototype testing. Constr Build Mater 368:130345. https://doi.org/10.1016/j.conbuildmat.2023.130345

Moore MM, Phares B, Graybeal B, Rolander D, Washer G (2001) Reliability of visual inspection for highway bridges, volume I: final report no. FHWA-RD-01-020. USDOT FHWA

NACE 01102 (2002) State-of-the-art report: criteria for cathodic protection of Prestressed concrete structures. AMPP, Houston, USA

NACE 01104 (2020) State-of-the-art report: electrochemical re-alkalization of steel-reinforced concrete. AMPP, Houston, USA

Office of Bridges and Structures (2018) Designing and detailing post-tensioned bridges to accommodate nondestructive evaluation. Report No. FHWA-HIF-18-029. USDOT FHWA

Patil S, Torres E, Freeman C, Hamilton HR (2019) Evaluation of drying technique for remediating soft grout in posttensioning tendons. PTI J:11–23

Patil SR, Torres E, Hamilton HR (2020) Evaluation of techniques to remove defective grout from post-tensioning tendons. Final report. University of Florida, USA

Pedeferri P (1996) Cathodic protection and cathodic prevention. Constr Build Mater 10:391–402. https://doi.org/10.1016/0950-0618(95)00017-8

Permeh S, Vigneshwaran KKK, Echeverria M, Lau K, Lasa I (2018) Corrosion of post-tensioned tendons with deficient grout, part 2: segregated grout with elevated Sulfate content. Corrosion 74:456–467. https://doi.org/10.5006/2568

Permeh S, Vigneshwaran KKK, Lau K, Lasa I (2019) Corrosion of post-tensioned tendons with deficient grout, part 3: segregated grout with elevated sulfate and vestigial chloride content. Corrosion 75:848–864. https://doi.org/10.5006/3097

Pillai RG (2009) Electrochemical characterization and time-variant structural reliability assessment of post-tensioned, segmental concrete bridges. Texas A&M University, Dissertation

Powers RG, Sagues AA, Virmani YP (2016) Corrosion of post-tensioned tendons in Florida bridges. Research Report No. FL/DOT/SMO/04–475, FLDOT

PTI (2003) Specification for grouting of post-tensioned structures. Post-Tensioning Institute, Phoenix, USA

Quaglini V, Pettorruso C, Cattaneo S, Rossi D (2023) Assessment of bridge post-tensioning systems using non-destructive (ND) inspection methods. ce/papers 6:9–17. https://doi.org/10.1002/cepa.2065

Rapaport G (2022) Evaluation of post tensioned bridges' tendon ducts by NDT and minor invasive measures. In: Paper presented at the international symposium on nondestructive testing in civil engineering, Zurich

Reis RA (2007) Corrosion evaluation and tensile results of selected post-tensioning strands at the SFOBB skyway seismic replacement project phase III report. Materials Engineering and Testing Services, Office of Testing and Technology Services, Corrosion Technology Branch, California

Rossi D, Pettorruso C, Bruschi E, Zoccolini L, Quaglini V (2024) Evaluation of non-destructive techniques for inspection of post-tensioned concrete bridges. Procedia Struct Integrity 62:307–314. https://doi.org/10.1016/j.prostr.2024.09.046

Sagues A, Powers GR, Wang H (2003) Mechanism of corrosion of steel strands in post-tensioned grouted assemblies. In: Paper presented at the CORROSION 2003, San Diego, California. Paper No. 03312, NACE International

Sagüés AA, Powers RG (2006) Mechanism and detection of corrosion of post-tensioned tendons in Florida bridges. In: PTI technical conference and exhibition, Reno, USA

Salas RM, Schokker AJ, West JS, Breen JE, Kreger ME (2004) Conclusions, recommendations and design guidelines for corrosion protection of post-tensioned bridges. Report No. FHWA/TX-04/0–1405-9, USDOT FHWA

Sharma S, Osmolska MJ, Hornbostel K, Bathen L, Novakova I, Geiker M (2024) Repair options for corrosion-damaged Prestressed concrete bridges. Nordic Concrete Res 70:99–123. https://doi.org/10.2478/ncr-2024-0004

Sprinkel MM (2015) VDOT experience with grouts and grouted post-tensioned tendons. PTI J Case Stud:51–61

Taveira LV, Sagues AA, Sabando JL (2008) Detection of corrosion of post-tensioned strands in grouted assemblies. In: Paper presented at the CORROSION 2008, New Orleans, Louisiana. Paper No. 08398, NACE International

Theryo TS, Hartt WH, Paczkowski P (2013) Guidelines for sampling, assessing, and restoring defective grout in prestressed concrete bridge post-tensioning ducts. Report No. FHWA-HRT-13-028. USDOT FHWA

Trejo D, Hueste MBD, Gardoni P, Pillai RG, Reinschmidt K, Im SB, Kataria S, Hurlebaus S, Gamble M, Ngo TT (2009a) Effect of voids in grouted, post-tensioned concrete bridge construction: volume 1—electrochemical testing and reliability assessment. Report No. FHWA/TX-09/0–4588-1 Vol. 1, Texas Transportation Institute, Texas

Trejo D, Hueste MBD, Gardoni P, Pillai RG, Reinschmidt K, Im SB, Kataria S, Hurlebaus S, Gamble M, Ngo TT (2009b) Effect of voids in grouted, post-tensioned concrete bridge construction: volume 2-inspection, repair, materials, and risks. Report No. FHWA/TX-09/0–4588-1 Vol. 2. Texas Transportation Institute, USA

Trejo D, Pillai RG, Hueste MBD, Reinschmidt KF, Gardoni P (2009c) Parameters influencing corrosion and tension capacity of post-tensioning strands. ACI Mater J 106:144–153. https://doi.org/10.14359/56461

Vigneshwaran KKK, Permeh S, Echeverria M, Lau K, Lasa I (2018) Corrosion of post-tensioned tendons with deficient grout, part 1: electrochemical behavior of steel in alkaline sulfate solutions. Corrosion 74:362–371. https://doi.org/10.5006/2541

Wang H, Sagues AA, Powers R (2005) Corrosion of the strand-anchorage system in post-tensioned grouted assemblies. In: Paper presented at the CORROSION 2005, Houston, Texas. Paper No. 05266, NACE International

Whitmore D, Lasa I (2018) Impregnation technique provides corrosion protection to grouted post-tensioning tendons. MATEC Web Conf 199:1–6. https://doi.org/10.1051/matecconf/201819905008

Yong-Sik Y, Hwa-Sung R, Hee-Seob L, Kyung-Taek K, Jeong-Su K, Seung-Jun K (2018) Effect of grout conditions and tendon location on corrosion pattern in PS tendon in grout. Constr Build Mater 186:1005–1015. https://doi.org/10.1016/j.conbuildmat.2018.08.023

Glass Fiber Reinforced Polymer (GFRP) Reinforcement in Concrete: Material Properties, Bond Behavior, Structural Performance, and Elevated Temperature Effects

M. Preethi, R. Kondalraj, and Chandru Pichaimuthu

Abstract Glass fibre-reinforced polymer (GFRP) bars provide a durable, corrosion-resistant alternative to steel reinforcement, particularly in aggressive environments. Their high tensile strength and lightweight nature enhance structural efficiency, yet their lower elastic modulus and brittle behaviour shift design emphasis toward serviceability, particularly in controlling crack width and deflection. Bond performance depends strongly on surface configuration, bar diameter, embedment length, and concrete confinement, with enhanced surface treatments improving mechanical interlock. Although GFRP exhibits excellent corrosion resistance, environmental factors such as high alkalinity, moisture ingress, freeze–thaw cycles, fatigue loading, and elevated temperatures may affect long-term performance. Hybrid reinforcement systems and fibre-enhanced concrete can improve structural response. Overall, effective application of GFRP requires serviceability-driven design and adherence to established code provisions to ensure reliable and sustainable infrastructure performance.

Keywords GFRP · Flexural behaviour · Shear performance · Surface characteristics

M. Preethi · R. Kondalraj (✉) · C. Pichaimuthu
Department of Civil Engineering, National Institute of Technology Calicut, Calicut, Kerala, India
e-mail: kondalrajr@nitc.ac.in

M. Preethi
e-mail: preethi_p250154ce@nitc.ac.in

C. Pichaimuthu
e-mail: chandru@nitc.ac.in

João M. P. Q. Delgado (ed.), *Advanced Materials, Corrosion Science, and Rehabilitation of Concrete Infrastructure*, Building Pathology and Rehabilitation 29,
https://doi.org/10.1007/978-3-032-28851-6_4

1 Introduction

Glass Fibre-Reinforced Polymer (GFRP) is an effective alternative to traditional steel reinforcement due to its corrosion-resistant properties and durability. They are composed of high-strength glass fibres embedded in a thermoset polymer matrix. It has been proven that GFRP bars are composed of approximately 75–80% glass fibre and 25% polymers. Polyvinyl esters, polyolefins, and epoxies are commonly used polymer matrices. GRPR has less carbon emissions about 17–66.8% than traditional steel and concrete. As it is more durable, it reduces the need of having frequent replacements, supporting sustainable construction practices. GFRP's lightweight nature lowers transportation costs and handling requirements. This efficiency may be particularly beneficial for remote or hard-to-access construction sites, especially in rural areas. GFRP works well with traditional construction materials, enabling seamless integration into existing or mixed-material frameworks. This versatility makes it the best and preferred choice for retrofitting or expanding existing infrastructure.

GFRP bars are manufactured by the Pultrusion Process. This process aligns the fibres longitudinally for optimal uniaxial performance. GFRP bars usually have a tensile strength in the range of 600–1000 MPa and weigh only 25% of the weight of conventional steel bars. This means that while steel can deform, GFRP rebar and mesh maintain their structural integrity without yielding, making them highly resistant to permanent deformation. Their non-conductivity suits MRI facilities, rail electrification, and airport runways. They can also be used in military structures that are undetectable by radar. Their non-corrosive properties offer a sustainable solution for concrete production, such as using seawater instead of freshwater directly in the concrete mix. Even though GFRP is a good alternative, it has many disadvantages.

1. When GFRP is cut in the field, the glass fibres inside it come out, allowing water to enter and weaken the bar over time.
2. Steel can be easily bend into any shape. But GFRP is brittle so cannot mould easily
3. When concrete is poured, GFRP is so light that it can float up and lose its cover position.
4. GFRP shows linear-elastic behaviour when it breaks suddenly, while steel has a ductile plateau

A comprehensive assessment of the physical properties, material optimisation, durability performance, and structural behaviour of concrete members reinforced with GFRP will provide an overall picture of the use of GFRP bars in concrete structures instead of conventional steel reinforcement. Due to its lower elastic modulus, the design of GFRP-reinforced concrete structures is governed primarily by serviceability limits, especially crack width and deflection. Concrete members reinforced with GFRP bars, when subjected to bending moments, exhibit linear behaviour up to cracking, after which the stiffness is reduced. It is found that deflections and strains of concrete members reinforced with FRP rods are generally larger than those reinforced with traditional mild steel rods. This will be due to the FRP reinforcement's

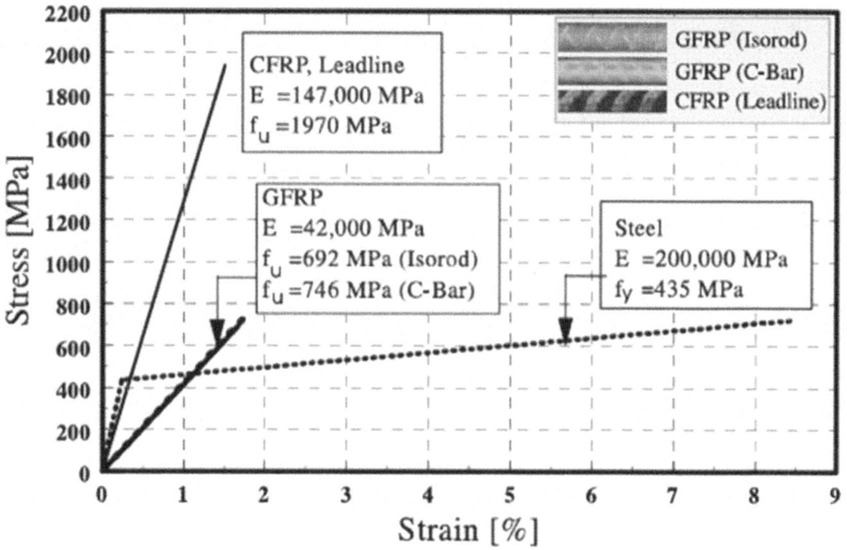

Fig. 1 Material characteristics of FRP and steel (Abdalla 2002)

lower elastic modulus and different bond characteristics (Abdalla 2002). Thus, it requires increased safety considerations and designs that prioritise serviceability. ACI 440 prescribes a serviceability factor of 2.5 for GFRP, while for steel it is just 1.67. A comparison of the material characterisation is given in Fig. 1. It gives a comparative study of GFRP, CFRP and traditional steel reinforcement.

Geometry optimisation of Glass Fibre Reinforced Polymer (GFRP) rebars focuses on improving the mechanical bond between the bar and concrete, enhancing structural efficiency, and tailoring shapes for specific applications. Bond performance between FRP bars and concrete can be analysed using pull-out or beam-end tests. The durability performance of GFRP bars can be improved by selecting appropriate constituents. Alkali-resistant fibres, when properly sized, may further enhance the durability of GFRP bars in harsh alkaline environments. The quality of the constituent materials of FRP bars, the manufacturing procedure, and the fibre/matrix influence the durability of FRP bars (Chen et al. 2007).

In general, the conventional GFRP has surface threads with a spacing equal to 0.8 times the bar diameter and a thread depth of 1 mm, but it has a relatively small core diameter. When used in reinforced concrete, GFRP bars develop only 60% of the bond strength of steel bars, meaning they grip the concrete less strongly and tend to slip more before failure. In a four-point bending test on beams reinforced with either GFRP or steel, both types of beams exhibited similar ultimate loads when the reinforcement ratio and beam size were the same. This means that their overall load-carrying capacity is comparable. However, GFRP bars are less effective than steel in controlling cracks because GFRP is less stiff and stretches more under load. As a result, beams reinforced with GPRP tend to develop wider cracks and experience greater slip, even though they can ultimately sustain loads up to those of steel-reinforced beams.

Several international guidelines govern the design and construction of concrete structures reinforced with Fibre-Reinforced Polymer (FRP) bars, including GFRP.

- JSCE 1997
- CSA S806-12
- ACI440.1R-15

2 GFRP Physical Classification and Material Optimisation

The physical properties of the new generation of GFRP bars are entirely different from the bars that were developed when ASTM D7957 was first published. Due to enhanced manufacturing processes, improved properties of GFRP bar constituents, and higher glass fibre mass fractions, the new generation of GFRP bars exhibits higher physical and mechanical properties. GFRP bars with a modulus of elasticity greater than 60GPa are available today. GFRP bars can be physically classified as follows:

1. Classification According to Structural Grade (CSA S807)
   (a) Grade I—standard modulus
   (b) Grade II—high modulus
2. Classification According to Durability (CSA S807)
   (a) D1 Classification: Standard durability for concrete exposure
   (b) D2/D3: Higher durability grades (used for alkaline environment)
3. Classification According to Fibre Type
   (a) E-Glass (Electrical Grade)
   (b) E-CR Glass (Electrical/Chemical Glass)
   (c) S-Glass/S2- Glass
4. Classification According to Manufacturing Process
   (a) Pultrusion
   (b) Strand/Winding
   (c) 3D braiding
5. Classification According to Surface Profile
   (a) Deformed/ribbed bars
   (b) helically deformed bars
   (c) helically grooved bars
   (d) double helical wrap/sand-coated
   (e) sand-coated bars

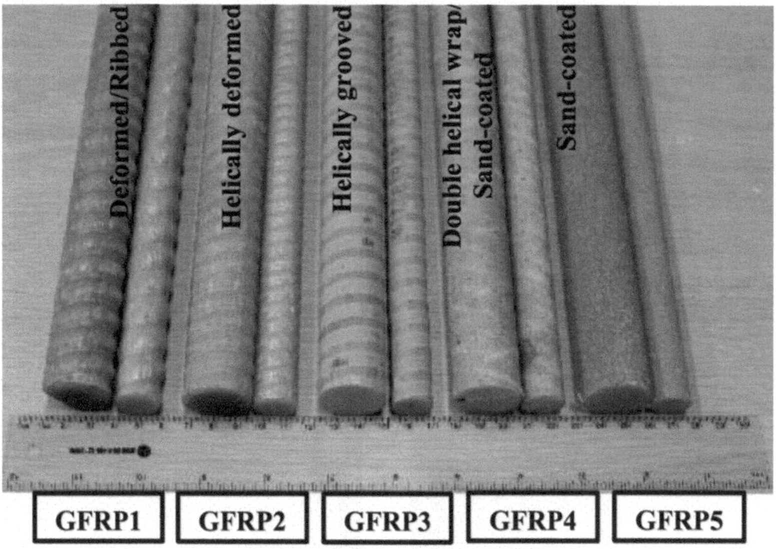

Fig. 2 GFRP bars types and surface characteristics (Benmokrane et al. 2023)

The commonly found surface types listed here are shown in Fig. 2.

From the above list, usually, GFRP bars are manufactured by the pultrusion process using corrosion-resistant E-glass, in which glass fibres are impregnated with a vinyl-ester resin in accordance with ASTM D7957-22 and CSA S807-19. The actual manufacturing process is shown in Fig. 3. GFRP bars made with thermoplastic resins are also advantageous over those made with thermosetting resins, as they are manufactured similarly to conventional steel bars. Thermoplastics are more rigid, have greater impact resistance, and are more deformable than thermosets (Benmokrane et al. 2023). However, due to irregular surface characteristics, there will be a difference between the measured and nominal cross-sectional areas, which could affect constructability using GFRP bars. The clear spacing restrictions in ACI 440.11-22 should be adjusted to account for the discrepancy in nominal and measured bar dimensions, or the maximum cross-sectional areas specified in ASTM D7957-22 should be tightened (El-Nemr et al. 2023). But studies by Yan et al. (2016) concluded that there is no exact correlation between failure modes and bar diameter, even though splitting failure and pullout remain the major failure modes analysed in terms of bar diameter. The bond strength decreased as the bar diameter increased in both cases.

2.1 Cross-Sectional Area and Diameter

Smaller-diameter bars generally develop higher average bond strength (τ) because the bond stress is distributed over a smaller perimeter, leading to more uniform

Fig. 3 Open mould pultrusion process (Omar Alajarmeh et al. 2025)

stress transfer along the embedment length. Larger bars tend to show reduced bond strength per unit area due to non-uniform stress distribution and greater internal slip. Additionally, larger bars require longer development lengths to achieve full tensile capacity. According to Rahman (2025) for GFRP bars specifically, the effect of diameter is also linked to their lower modulus of elasticity compared to steel. Larger GFRP bars may exhibit greater slip and wider cracks under service loads due to reduced stiffness. Therefore, increasing bar diameter typically results in:

- Lower average bond stress

- Higher risk of concrete splitting
- Increased required development length
- Greater sensitivity to concrete cover and confinement

The study by Luo et al. (2024) explains how different geometric properties of GFRP bars influence their bond strength with Ultra High-Performance Concrete (UHPC). The researchers found that the relationship between bar diameter and bond strength is non-monotonic, meaning that it does not simply increase or decrease continuously. When the bar diameter increased from 8 to 18 mm, the bond strength improved because a larger diameter provided more surface area and better mechanical interlocking with the concrete. However, as the bar diameter increases further, bond strength decreases. This may be due to changes in stress distribution and reduced confinement efficiency around larger bars, which weakens the bond performance. The Bond stress relationship of GFRP bars in UHPC is shown in Fig. 4.

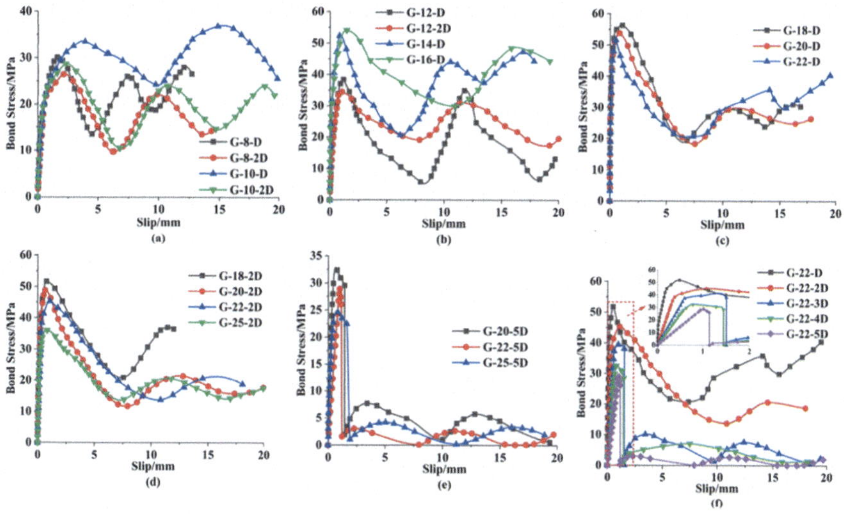

Fig. 4 Bond stress-slip response of GFRP bar embedded in UHPC

3 Durability and Corrosion Resistance

3.1 Exposure to Various Environmental Factors

3.1.1 Exposure to Alkaline Environment

Concrete made with Fresh water (FW) is usually stronger than concrete made with seawater (SW) at the beginning. But both types of concrete become stronger over time, especially when kept at higher temperatures. However, in GFRP-reinforced concrete, the bond strength was initially lower in seawater concrete; in fact, it was about 43% lower than in freshwater concrete at the early stage. However, as concrete ages, the difference becomes smaller. Over time, both FW and SW concrete exhibit similar bond strengths. After two years of immersion in seawater, the bond strength increases. It is found that about a 24% increase for freshwater concrete at 20 °C and about a 65% increase for seawater concrete at 60 °C. Also, the studies highlight that the specimens have superior compressive strength during the initial stages under freshwater conditions. Over time, it will attain strength in both freshwater and seawater. But the transverse shear strength is found to be decreased, which may be due to the debonding of the fibre (Nepomuceno et al. 2025). At the same time, according to Sun et al. (2023), a decrease in tensile strength is observed after 300 days of exposure to an alkaline environment of pH 10, which was approximately 1.23%, while the maximum percentage loss of Interlaminar shear strength (ILSS) was approximately 29%. However, a further increase in pH led to a significant decrease in tensile and shear strength. The experiment showed that the deterioration of the mechanical properties of GFRP bars in alkaline solutions cannot be determined solely by pH, and the actual factor causing the performance degradation was COH, which was linearly related to the rate of Tensile Strength loss. This was analysed through an accelerated ageing test conducted for 12 months at 60 °C. The results of this study provide engineers with a detailed database on the degradation of GFRP tensile strength, indicating that reducing the exposure pH, rather than the saline content, is essential for improving GFRP durability (Wang et al. 2023).

3.1.2 Freeze-Thaw Cycles

Fursa (2016) developed a non-destructive testing method for determining the bond behaviour of GFRP in concrete. Theoretical and experimental studies to assess the durability of glass fibre-reinforced polymer (GFRP) rebar in concrete under cyclic freeze-thaw conditions by measuring electrical response to mechanical impact. Here, time and frequency-related features of the measured signals are determined. The results showed that cyclic freeze-thaw action weakens the rebar-concrete bond, and after 18 cycles, the bond strength is reduced by 2 times. These results can be verified by observing Alachek et al. (2018), who suggested that failure due to freeze-thaw has shifted to adhesive failure of the GFRP bar, with combined interfacial failure of the concrete-GFRP bond. For cold-weather applications, GFRP bars in engineered cementitious composites (ECC) last longer than steel reinforcement in regular concrete when it comes to bond endurance under freeze-thaw (FT) cycles. After 150 FT cycles, the average bond strengths declined by 88.41% in 30 MPa regular concrete, 38.86% in 30 MPa ECC, and 58.21% in 50 MPa ECC. ECC's fibre bridging and micro-crack management (strain-hardening) kept the interface intact better than fragile conventional concrete, even though it was stronger (Wang 2024). Most of the free-thaw degradation is by fibre resin debonding.

3.1.3 Fatigue Behaviour

Studies show that GFRP lasts 20 times longer under fatigue loading than steel, with 24% less cumulative damage at high stress levels. However, it is less rigid (29% less), which means it bends more easily. This makes it perfect for bridges and other places where corrosion is a problem. For the first 90% of their fatigue life, hybrid GFRP/Steel-RC beams maintain stable strains in the GFRP bars. However, in the last 10%, they drop off quickly because of cumulative cracking and interface slip. Higher reinforcement ratios increase fatigue life by better spreading cyclic stresses across fibres. Lower stress levels lengthen cycles by slowing down the buildup of strain in the low-modulus GFRP (Zhu et al. 2017). GFRP-RC beams bend more (14 mm vs. 6 mm at failure) and lose stiffness faster when static loads are applied because GFRP has a low modulus. This makes cracks grow wider and faster than Steel-RC.GFRP-RC has lower dynamic stiffness at the initial stages, but it maintains higher residual stiffness after each cycle, indicating 24% less damage. Steel, on the other hand, gets more damage from cyclic yielding/plasticity. Hybrid GFRP/Steel rebar strikes a balance: steel provides early stiffness and crack resistance, while GFRP increases long-term fatigue resistance without the risk of corrosion (Wu et al. 2023).

When subjected to cyclic loading, GFRP-RC exhibits lower initial dynamic stiffness compared to Steel-RC structures yet demonstrates greater retention of residual stiffness after several cycles, characterised by reduced deterioration and an extended fatigue life. Both get more damage from fatigue at lower stress levels. This makes them last longer, but it also adds up to more damage over time. GFRP is more stable because it doesn't corrode from fatigue.

3.1.4 Moisture Absorption

Moisture absorption is another major problem associated with GFRP materials. Because it affects the polymer matrix of GFRP, which will impact the electrical, mechanical and dimensional characteristics of the product. Different degradation patterns were observed due to moisture exposure in the glass fibre, interphase, and resin matrix. The glass fibres and the resin matrix showed the least degradation, followed by the interphase. Because of the increased concrete porosity for moisture transfer, higher w/c ratios accelerated degradation (Liew et al. 2025). When FRP rebars are kept in a very humid environment for many years, they slowly absorb moisture. The water enters faster along the fibre direction than across it. Short samples absorb moisture quickly, but long rebars take several years to fully absorb water. Because of this moisture, the strength between the layers of the material (interlaminar shear strength) decreases. Especially in aramid FRP, while carbon FRP is affected the least. Most of the strength loss happens in the first few years, and after that, it does not reduce much further. After drying, about 90% of the sample's strength is recovered, indicating that most of the damage is not permanent (Starkova et al. 2021).

3.1.5 Exposure to Elevated Temperature

The steel and concrete have almost the same coefficient of thermal expansion. So that the stresses formed during expansion will be negligible or zero. But the thermal expansion coefficient of the fibre is entirely different from that of concrete, and it is much higher (Balázs 2001). So that this difference may induce elevated stresses during expansion when exposed to greater temperature variation, and may, in turn, affect bond performance (Ceroni 2006). The tensile performance of GFRP bars is significantly affected even when embedded in concrete with cover at elevated temperatures. Still, the concrete cover behaves as a protective barrier at moderate temperatures. The difference is evident in Fig. 5. Below the glass transition temperature, the tensile strength and stiffness remain almost unchanged, whereas between the glass transition temperature and the resin decomposition temperature, the tensile strength gradually decreases due to softening and weakening of the resin matrix. The resin matrix undergoes significant thermal breakdown above its decomposition temperature (around 350–400 °C), resulting in a sudden and irreversible loss of strain capacity, elastic modulus, and tensile strength. The concrete cover delays the heat transfer towards the GFRP rebars, resulting in better strength retention at temperatures below 400 °C. However, the protective function of the concrete layer is substantially lost when temperatures rise beyond 400 °C, and it begins to crack severely. Then both embedded and naked rebars have comparably low residual strength. An enhanced prediction model was developed to more precisely estimate the residual tensile strength of GFRP rebars embedded in concrete following fire exposure, accounting for heat absorption and temperature lag effects. This model incorporated a Heat Absorption Parameter to analyse the amount of heat the concrete absorbs before reaching the embedded GFRP bar during fire exposure, and a Strength Retention Difference Parameter to determine the difference in tensile strength between protected and unprotected rebar (Qian et al. 2025).

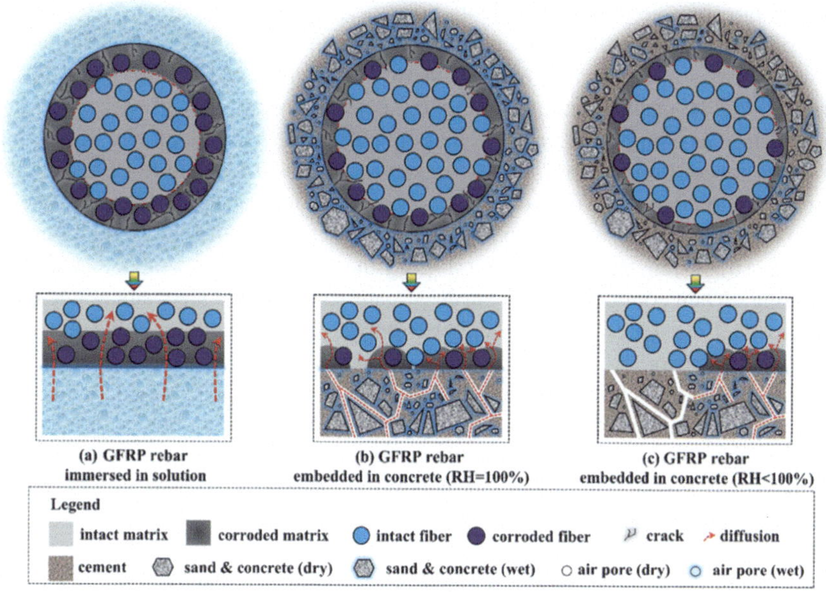

Fig. 5 States of bar: **a** immersed in solution, **b** embedded in concrete, (RH = 100%) **c** embedded in concrete (RH < 100%)

4 Serviceability Factors and Structural Performance

4.1 Bond Performance

Bond performance of GFRP in concrete refers to the bar's ability to transfer tensile forces to the surrounding concrete through various mechanisms. Generally, the bond of the reinforcing bar to concrete can be indicated as

(i) Adhesion resistance of the interface between bar and concrete, defined as a chemical bond
(ii) Frictional resistance of the interface against slip and
(iii) Mechanical interlock formed due to irregularity of the interface.

Benmokrane et al. (2023) have conducted Pullout tests on GFRP bars with varying surface characteristics to assess bond strength. The pullout specimen constructed is given in Fig. 6. The bonded length was kept constant at five times the bar diameter ($5d_b$), where d_b is the nominal diameter of the bar. They have conducted the experiment using the Baldwin loading machine, with the load applied at a controlled rate of 0.1 kN/min to ensure accurate measurement of bond behaviour. They chose the specimen with normal-strength concrete of 35 MPa. However, the compression strength test using a cylindrical specimen showed a strength of 41.6 MPa on the day of testing. This confirms that the concrete achieved slightly higher strength than the target value, ensuring reliable bond performance evaluation. Also, it was observed that the deeper the grooves on GFRP bars, the higher the bond performance. This will be due to greater mechanical interlocking between the bar and the surface concrete. While tensile force is applied, the concrete must resist greater shear and bearing stresses along these grooves, making pull-out more difficult and improving bond resistance. Furthermore, bars with surfaces created by double helical wrapping and sand coating exhibited stronger bonds. By increasing surface roughness, the sand coating increases friction between the concrete and the bar. Better confinement and stress transfer are made possible by the double helical wrap's mechanical anchoring. When combined, these surface treatments enhance mechanical interlock and frictional resistance, resulting in better bond performance than with smooth or single-textured bars.

Also, Luo et al. (2024) showed that relative rib height (rib height divided by bar diameter) was among the most significant geometric parameters influencing bond strength. Greater resistance to pull-out results from improved mechanical interlocking between the bar and UHPC caused by higher relative rib height. Furthermore, though to a lesser degree, two additional factors also affect bond behaviour: the rib inclination tangent, which represents the rib angle, and the rib width ratio, which is the ratio of rib width to spacing. These factors influence how well the ribs resist slippage between the bar and concrete and transfer shear forces.

El-Nemr et al. (2023) investigated the bond behaviour of FRP bars using 87 pullout specimens, considering the parameters such as bar diameter, fibre type, surface configuration, modulus of elasticity and concrete strength. Deformed steel bars of size 10 and 15 mm were also tested as control specimens for comparison. The

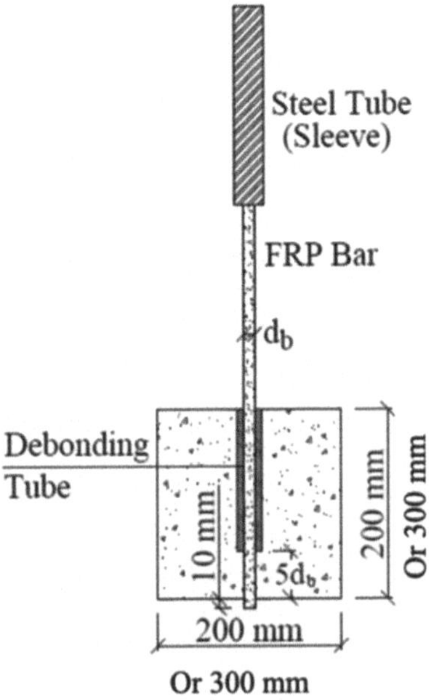

Fig. 6 Bond test specimen (El-Nemr et al. 2023)

GFRP bar diameters ranged from 13 to 25.1 mm and were embedded in either normal-strength concrete of 35 MPa or high-strength concrete of 65 MPa. The results showed that bond strength is strongly influenced by the parameters listed above. Increasing concrete strength improves bond performance because stronger concrete provides better confinement and resistance to slipping. However, increasing bar diameter reduces bond strength due to higher stress concentration and reduced confinement efficiency. Here, the sand-coated bar performed better because the rough surface enhances friction and mechanical interlocking. Although FRP bars exhibited lower bond strength than steel bars, their performance still met code requirements. The mode of pullout failure depended on the surface configuration and concrete strength: in steel bars, failure occurred due to crushing of concrete around the ribs; in FRP bars embedded in normal strength concrete, failure was caused by crushing or splitting of the wrapped and sand-coating layer; and in high strength concrete, failure occurred due to peeling off of the sand-coating or grooved layer. The test results from the experiment are depicted in the chart given in Fig. 7.

Aly et al. (2007) found that increasing the concrete cover significantly improves the bond strength between GFRP bars and concrete. Specifically, when the concrete cover was increased to four times the bar diameter, the bond strength increased by about 27%. This happens because a thicker concrete cover provides better confinement around the GFRP bar. Confinement means the surrounding concrete resists the radial tensile stresses that develop when the bar is pulled, allowing the bar to generate higher forces against the concrete surface. As a result, the bond between the

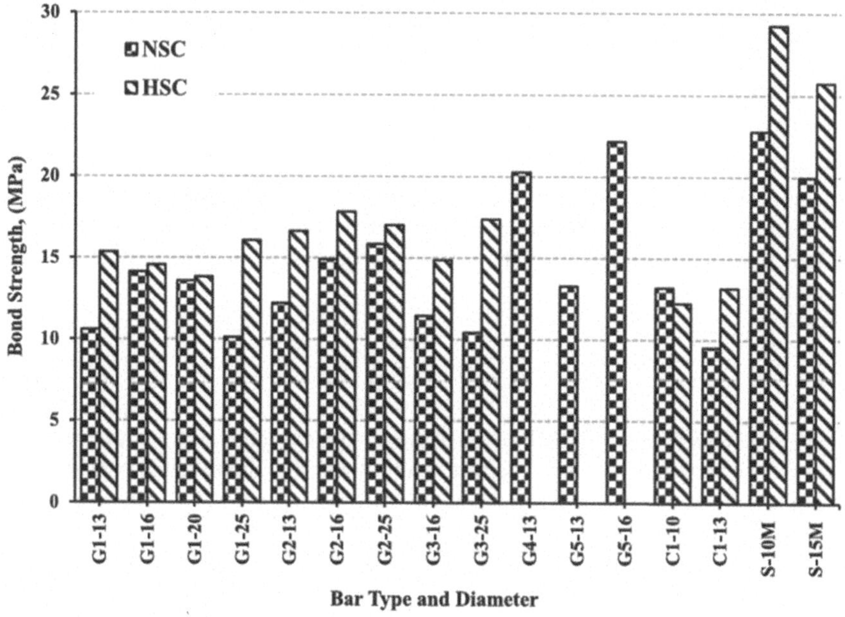

Fig. 7 Bond performance of GFRP when embedded in NSC and HSC

bar and concrete becomes stronger. Similarly, Alves et al. (2010) observed a change in failure mode with varying concrete cover thickness. When the concrete cover was equal to one time the bar diameter, splitting failure occurred. Splitting failure occurs when tensile stresses around the bar cause cracks to form and propagate through the concrete cover, leading to sudden failure. However, when the concrete cover was increased to at least twice the bar diameter, the failure mode shifted to pullout failure or even bar rupture. This means the concrete was strong enough to prevent premature cracking, forcing the bar to either slip out gradually or reach its tensile capacity and fracture.

Embedment length is another critical parameter that influences the bond strength between the FRP bar and concrete. Generally, an embedment length of five times the bar diameter is provided, which gives a desirable bond-pullout failure for GFRP bar-to-concrete. It is found that the maximum average bond stress of GFRP bars to concrete decreases as the embedment length increases. It may be due to a nonlinear distribution of the bond stress along the reinforcing bar (Fava et al. 2016). Also, Achillides et al. (2004) reported that increasing embedment length not only decreases the maximum average bond stress of FRP bars but also yields a lower initial bond stiffness, corresponding to a high nonlinear, non-uniform distribution of bond stress along the longer bar.

4.2 *Tensile Strength*

Standard GFRP rebars, which are commercially available today, generally exhibit tensile strengths of 483-1600 MPa, with design strengths often exceeding 1000 MPa. Benmokrane et al. (2023) has conducted tensile tests of GFRP bars of size 15.9 mm and 25.4 mm nominal diameters, and the samples were attached with a 200 mm high-accuracy linear variable differential transformer (LVDT) to obtain the sample elongation in the middle part of the sample in between the grips during testing in accordance with ASTM D7205-21. A Baldwin testing machine was used to test the bars up to failure. They used the following equation to find the ultimate tensile strength *(fu)*.

$$f_u = \frac{F_u}{A}$$

Where f_u is tensile strength (MPa); F_u is the tensile capacity, and A is the nominal cross-sectional area of the bar. The test results show that the selected bars met the ASTM requirements. The test results and the ASTM and CSA (2017) requirements are presented in Table 1.

Table 1 ASTM and CSA requirements of GFRP bars

Property	ASTM D7957-22	CSA S807-19	ASTM D7957-22	CSA S807-19
Ultimate load (kN)	≥ 130	–	≥ 297	–
Tensile strength (MPa)	–	≥ 1000	–	≥ 1000
Tensile modulus (GPa)	≥ 44.8	≥ 60.0	≥ 44.8	≥ 60.0
Ultimate strain (%)	≥ 1.1	≥ 1.2	≥ 1.1	≥ 1.2

4.3 *Flexural Behaviour*

Flexural behaviour of GFRP-reinforced structures is most assessed using three-point or four-point bending tests. An actuator is used for applying a load in displacement control mode. Displacement and strain are obtained using Linear Variable Differential Transformers and Digital Image Correlation techniques. Since GFRP material has a lower elastic modulus, the design considerations mainly focus on serviceability performance, especially crack width and deflection. Various studies are ongoing on flexural behaviour and serviceability limits of GFRP. Adam et al. (2015) has conducted a study to find the flexural behaviour of commercially available GFRP bars. They have conducted both experimental and analytical studies to assess the behaviour. The test results showed that both crack width and mid-span deflection have reduced significantly as the reinforcement ratio has increased. The ultimate load for the highest reinforcement ratio has increased by 97%, indicating some ductility similar to that of the concrete. Also, it was concluded that, as concrete tensile strength is somewhat related to its compressive strength, the cracking load can be improved by using higher-grade concrete mixes. The indication of structural capacity governed by a higher reinforcement ratio can also be identified from the work of Ju et al. (2017). A direct proportionality can be identified between bar diameter and the effective compressive strength of concrete. The design equations recommended by ACI 440.1 R-15 (2015) for effective moment of inertia are based on Branson's equation. This can be identically used to calculate the deflection of the cracked section.

$$s_{\max} = 1.15\frac{E_f w}{f_{fs} k_b} - 2.5c_c \leq 0.92\frac{E_f w}{f_{fs} k_b}$$

From the equation, it is also clear that the bar spacing directly relates to the maximum allowable crack width. It is true that the bar spacing must be narrower to incorporate more flexural reinforcement to resist larger applied moments. But the bar spacing beyond a limit inversely affects the bond properties, and the crack width becomes conservative.

Many studies have been conducted on adding fibres to concrete to enhance its flexural behaviour. Steel and macro-synthetic fibres are among the most used types of fibres in structural applications. The tendency of steel to corrode may reduce the durability of concrete. Prolonged exposure to corrosive environments, such as salt solutions, can severely compromise the post-crack performance of steel FRC by

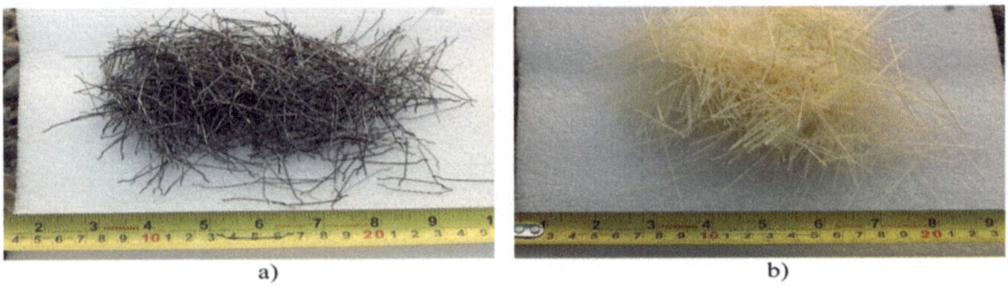

Fig. 8 Discrete fibres **a** Hooked end steel fibres **b** Polyolefin (PO) fibres (Balaji et. al. 2025)

corroding the fibres. Consequently, steel fibres might not be suitable in corrosion-prone environments such as outdoor or marine settings. According to ACI 544.4R-18, macro-synthetic fibres effectively improve serviceability and residual strength in concrete elements under flexure and shear, even though they are less stiff than steel fibres. This popularity is mainly due to their improved properties, such as ductility, corrosion resistance, resistance to alkali attack, and economy and durability (Richardson 2005). Furthermore, macro-synthetic fibres are lighter than steel fibres, making them an ideal choice when minimising the weight of construction materials is an important consideration, such as in modular construction.

Patil (2024) suggested that hybrid steel–polyolefin fibres in GFRP-reinforced concrete columns significantly enhanced compressive strength, energy absorption, post-peak response, and deformability under eccentric loading. Ibrahim et al. (2025) analysed that combining macro polypropylene fibres with micro basalt fibres results in better crack control than using macro fibres alone. At the same time, J. Wei et al. (2024) suggested that using Shrinkage Resisting Admixtures improved the performance of mixes by about 4–7% compared with mixtures made without SRA. This enhancement results from reduced surface tension in cementitious materials, thereby reducing the need for superplasticisers (Figs. 8 and 9).

A combination of polyolefin and steel fibres is used to enhance performance while reducing steel fibre corrosion. Polyolefin fibres have an elastic modulus of about 10 GPa, while steel fibres reach about 200 GPa. By blending flexible polyolefin fibres with stiffer steel fibres, the formulation aims to achieve a synergistic effect that enhances crack-arresting performance across different crack widths. Even the addition of polyolefin improves the behaviour, its concentration should be considered as an important factor. The researchers conclude that a percentage range of 0.7–1% shows major change in compressive strength of the concrete. At the same time, bond performance is analysed to predict serviceability (Balaji et al. 2025) (Fig. 10).

4.4 Shear Behaviour

Shear failure is sudden and brittle. It is found that shear failure is somewhat more dangerous than flexural failure and causes heavy damage. It occurs because principal

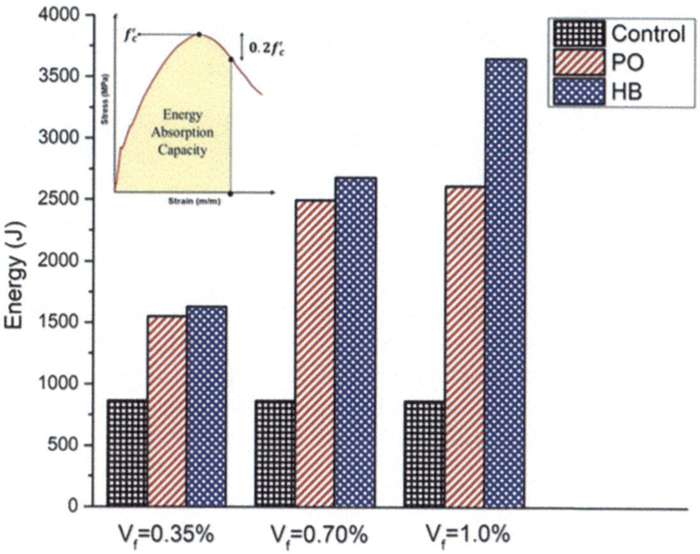

Fig. 9 Strain energy absorption capacity of PO and hybrid beams for various fibre dosages (Amal Dev et al. 2020)

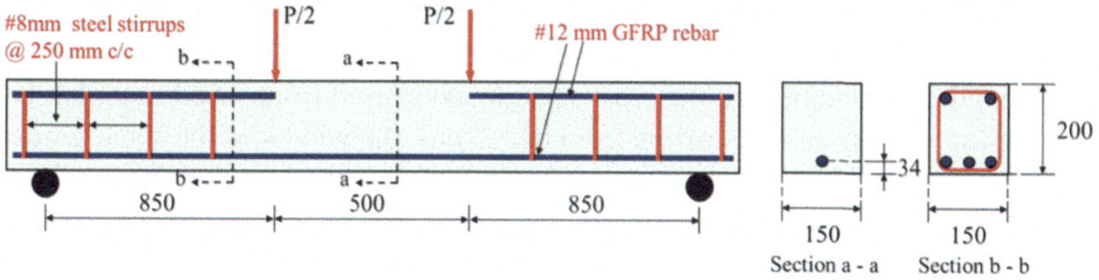

Fig. 10 Specimen details of four point bending test (Balaji et al. 2025)

tensile stresses exceed the ultimate tensile strength of concrete at the critical shear regions of the structure. We know that shear cracks propagate faster than flexural cracks. Xingyu Gu et al. (2016) found that the beam fails when a diagonal crack forms in the shear span, leading to shear failure as the crack rapidly propagates. This crack later extends horizontally at the level of GFRP bars, indicating bond failure due to slipping between the GFRP rebar and the surrounding concrete. In the case of flexural failure, the failure is found to be because the GFRP bar ruptured in tension either at the mid-span, where the bending moment is maximum or beneath the applied load point. Since GFRP does not yield like steel and behaves in a brittle manner, it suddenly fractures once its tensile strength is reached, leading to beam failure. It is therefore very important to check the shear capacity of the structure during design. Many studies are ongoing that focus on enhancing shear behaviour across various aspects. A normal provision of shear stirrups is depicted in Fig. 11. The application

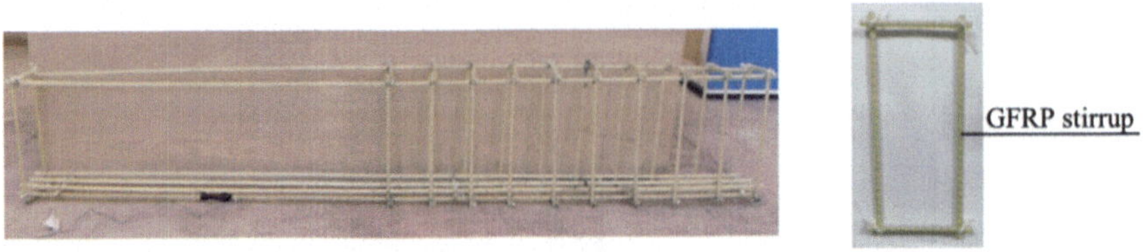

Fig. 11 Shear reinforcement detailing using GFRP

of prestressed concrete structures, the use of high-strength concrete, the addition of fibres to concrete, and the adoption of improved construction procedures can enhance shear performance (Zhang et al. 2025). Leopoldo Tesser (2013) analysed RC beams with internal steel trusses along with varying shear span to depth ratio. The experiments demonstrated that truss insertion effectively increases the structure's shear capacity.

Pre-built trusses made of 2 mm-thick welded steel strips in shear zones were compared with straight bars of the same size. Truss-strip beams were superior to bar-reinforced beams at controlling major crack width and ultimate shear capacity because they had better crack bridging and aggregate interlock (Etman et al. 2021). Monaco et al. (2014) used nonlinear finite element modelling to predict shear resistance and load transfer between steel truss members and concrete matrix. This allowed us to observe nonlinear behaviours such as cracking and bond sliding. Moreover, Jumaa et al. (2019) reported that when BFRP bars and stirrups were added to high-strength concrete beams, the crack spacing decreased from 0.52d to 0.36d. By changing the spacing of the stirrups from d/2 to d/3, the widths of the shear cracks got even less, which made the beams more usable, especially in smaller and medium-sized beams. As the transverse reinforcement ratio increased, the T-section beams with GFRP bars had more cracks, but they were smaller. Ultimate shear went up with the longitudinal reinforcement ratio because the dowel action and aggregate interlock were better, even though GFRP is less rigid (Debaiky and Kotynia 2020). Noor et al. (2012) concluded that the shear capacity of GFRP-longitudinally reinforced beams was closely related to the span/depth and stirrup spacing. A lower span/depth or tighter spacing increased the stresses. The cracks were closer together and more numerous than in steel RC beams, indicating GFRP's lower bond strength and flexibility.

Stirrups delayed but did not eliminate shear dominance: Flexural cracks extended farther toward mid-span, and diagonal cracks continued to propagate from the load point to the support. Closer spacing of about 150 mm probably caused cracks that were more common but thinner. The spalling on the cover near the support shows that the concrete is breaking down in a small area. This is made worse by GFRP's low elasticity, which means that stirrups crack before breaking, unlike steel, which bends. The final mode remains shear failure, but it has slightly greater capacity and ductility due to transverse restraint. Externally bonded CFRP sheets increased the shear capacity of GFRP-RC beams by 10% to 24%. However, ACI 440.2R overestimates these benefits because of debonding and internal stirrup interference. GFRP stirrups work like steel for shear across OPC, FRC, and GPC. Adding GFRP to geopolymer is a good green alternative that works well with OPC. Fibres make the structure more flexible, but they won't make it stronger. When CFRP covers the stirrups, the strains are balanced. However, when CFRP is offset, it carries more load, leading to increased deflection and interaction losses (Junaid et al. 2022).

The study by Faizan et al. (2023) concludes that anchored GFRP U-wraps are highly effective in enhancing shear performance both under normal and elevated temperatures.

Although they show a slight reduction in performance near the epoxy's glass transition temperature. While GFRP sheets exhibited significant decreases in tensile strength and stiffness around 55–60 °C, the strengthened beams still achieved up to 96% shear strength improvement at 60 °C compared to 112% at room temperature. Failure of wrapped beams was governed by anchor pullout and FRP debonding rather than rupture of the GFRP sheets, indicating that the full tensile capacity of the FRP was not utilised.

5 Conclusions

- Glass fibre reinforced polymer (GFRP) bars present a technically viable and durable alternative to conventional steel reinforcement in concrete structures, particularly in environments where corrosion resistance and long-term durability are critical. Owing to their high tensile strength-to-weight ratio, electromagnetic neutrality, and reduced maintenance requirements, GFRP reinforcements contribute to sustainable and resilient infrastructure systems. However, their lower elastic modulus and linear-elastic brittle behaviour shift structural design control from ultimate strength considerations to serviceability requirements, especially crack width limitation and deflection control.
- Bond performance between GFRP bars and concrete is strongly influenced by surface configuration, bar diameter, embedment length, and concrete confinement. Enhanced surface treatments such as helical deformation, sand coating, and double wrapping significantly improve mechanical interlock and frictional resistance. Nevertheless, bond characteristics remain sensitive to geometric parameters and environmental exposure.
- Durability assessments indicate that while GFRP bars exhibit excellent corrosion resistance, their mechanical properties may degrade under prolonged exposure to elevated temperatures, highly alkaline environments, freeze–thaw cycles, and moisture ingress. Proper material selection, optimised manufacturing processes, and adequate protective detailing are therefore essential to ensure long-term performance. The incorporation of hybrid reinforcement systems, macro-synthetic fibres, and shrinkage-reducing admixtures further enhances flexural, shear, and fatigue behaviour.

- Overall, successful implementation of GFRP-reinforced concrete systems requires adherence to established design guidelines and a serviceability-driven design philosophy. Continued research focused on long-term durability modelling, fire performance prediction, and structural optimisation will facilitate broader adoption and improved reliability of GFRP reinforcement in modern infrastructure applications.

References

ACI Committee 440 (2015) ACI 440.1R-15: guide for the design and construction of structural concrete reinforced with FRP bars. American Concrete Institute

Al-Zahrani MM (2024) Durability of concrete-embedded GFRP bars after 20 years of tidal zone exposure: correlation with accelerated aging tests. Case Stud Constr Mater 21:e03435. https://doi.org/10.1016/j.cscm.2024.e03435

Abbas H, Alanazi MA, Elsanadedy H, Almusallam T, Abadel A, Al-Salloum Y (2023) Effect of tie spacing and GFRP rebar diameter on compression behavior of concrete columns. J Build Eng 72:106740. https://doi.org/10.1016/j.jobe.2023.106740

Abdalla HA (2002) Evaluation of deflection in concrete members reinforced with fibre reinforced polymer (FRP) bars. Compos Struct 56(1):63–71. https://doi.org/10.1016/S0263-8223(01)00188-X

Achillides Z, Pilakoutas K (2004) Bond behavior of fiber reinforced polymer bars under direct pullout conditions. J Compos Constr 8(2):173–181. https://doi.org/10.1061/(ASCE)1090-0268(2004)8:2(173)

Adam MA, Said M, Mahmoud AA, Shanour AS (2015) Analytical and experimental flexural behavior of concrete beams reinforced with glass fiber reinforced polymer bars. Constr Build Mater 84:354–366. https://doi.org/10.1016/j.conbuildmat.2015.03.057

Alajarmeh O, Manalo A, Yatsenko D (2025) High elastic modulus GFRP rebar: a guide to manufacture and design of the influencing parameters. Composites Part C: Open Access 16:100576. https://doi.org/10.1016/j.jcomc.2025.100576

Alachek I, Reboul N, Jurkiewiez B (2018) Bond strength degradation of GFRP-concrete elements under aggressive exposure conditions. Constr Build Mater 179:512–525. https://doi.org/10.1016/j.conbuildmat.2018.05.249

Aly MS, Benmokrane B, El-Ragaby A (2007) Effect of concrete cover and transverse reinforcement on the bond behavior of FRP reinforcing bars. J Compos Constr 11(5):522–531

Alves JD, Branco FA, Ferreira JG (2010) Influence of concrete cover on the bond behaviour of GFRP bars embedded in concrete. Constr Build Mater 24(12):2450–2457

Balaji GR, Prakash SS (2025) Effect of discrete fiber addition on the serviceability performance of glass fiber reinforced polymer concrete beams under flexure. Structure 76:108901. https://doi.org/10.1016/j.istruc.2025.108901

Balázs GL, Borosnyói A (2001) Long-term behavior of FRP. In: Proceedings of the international workshop on composites in construction, Porto, Portugal, pp 175–184. https://doi.org/10.1061/40596(264)10

Benmokrane B, Mehany S, Shield C, Nanni A, Brown V (2023) Physical properties, longitudinal tensile properties, and bond strength of the new generation of GFRP bars. J Compos Constr 27(6):04023056

Bischoff PH, Gross SP (2011) Design approach for calculating deflection of FRP-reinforced concrete. J Compos Constr 15(4):490–499. https://doi.org/10.1061/(ASCE)CC.1943-5614.0000195

Cao Q, Jia Z, Zhou C (2025) Flexural and serviceability behaviour predictions of concrete beams reinforced with hybrid reinforcement of GFRP bars and highly ductile stainless steel bars. Structure 77:109102. https://doi.org/10.1016/j.istruc.2025.109102

CSA (2017) CSA S806-12: design and construction of building structures with fibre-reinforced polymer. Canadian Standards Association

Caratelli A, Meda A, Rinaldi Z, Spagnuolo S, Maddaluno G (2017) Optimization of GFRP reinforcement in precast segments for metro tunnel lining. Compos Struct 181:336–346. https://doi.org/10.1016/j.compstruct.2017.08.083

Ceroni F, Pecce M, Gaetano M, Cosenza A (2006) Bond behavior of FRP bar reinforced concrete members under elevated temperatures. J Compos Constr 10(1):43–51. https://doi.org/10.1061/(ASCE)1090-0268(2006)10:1(43)

Chen Y, Davalos JF, Ray I, Kim HY (2007) Accelerated aging tests for evaluation of durability performance of FRP reinforcing bars for concrete structures. Compos Struct 78:101–111. https://doi.org/10.1016/j.compstruct.2005.08.015

Debaiky D, Kotynia R (2020) Shear behaviour of concrete beams reinforced with GFRP bars: influence of longitudinal reinforcement ratio and shear span-to-depth ratio. Compos Struct 236:111878. https://doi.org/10.1016/j.compstruct.2019.111878

Dev A, Chellapandian M, Prakash SS (2020) Effect of macrosynthetic and hybrid fibers on shear behavior of concrete beams reinforced with GFRP bars. J Bridg Eng. https://doi.org/10.1061/(ASCE)BE.1943-5592.0001557

El-Nemr A, Ahmed EA, Benmokrane B (2023) Bond performance of fiber reinforced polymer bars in normal- and high-strength concrete. Constr Build Mater 393:131957. https://doi.org/10.1016/j.conbuildmat.2023.131957

Etman E, Ebid AA, El-Sayed MA, Reda Taha MM (2021) Shear behavior of reinforced concrete beams using welded steel strip trusses as shear reinforcement. Eng Struct 242:12543. https://doi.org/10.1016/j.engstruct.2021.112543

Fava G, D'Antino T, Carloni C (2016) The influence of bond length on the bond behavior of GFRP bars embedded in concrete. Compos Part B Eng 93:211–220. https://doi.org/10.1016/j.compositesb.2016.03.024

Fursa TV, Utsyn GE, Korzenok IN, Petrov MV, Reutov YA (2016) Using electric response to mechanical impact for evaluating the durability of the GFRP-concrete bond during the freeze-thaw process. Compos Part B Eng 90:392–398. https://doi.org/10.1016/j.compositesb.2015.11.026

Gu X, Zhang Y, Dai G (2016) Experimental investigation of the behaviour of concrete beams reinforced with GFRP bars under static and impact loading. Eng Struct 113:220–232. https://doi.org/10.1016/j.engstruct.2016.01.044

Hao Q, Wang Y, He Z, Ou J (2009) Bond strength of glass fiber reinforced polymer ribbed rebars in normal strength concrete. Constr Build Mater 23:865–871. https://doi.org/10.1016/j.conbuildmat.2008.04.011

Ibrahim E, Salama AE, Bakouregui AS, Yahia A, Benmokrane B (2025) Deflection and cracking of fiber-reinforced self-consolidated concrete beams reinforced with GFRP bars under cyclic loading. J Compos Constr. https://doi.org/10.1061/JCCOF2.CCENG-5049

JSCE (1997) Recommendation for design and construction of concrete structures using continuous fibre reinforcing materials. Japan Society of Civil Engineers

Ju M, Park Y, Park C (2017) Cracking control comparison in the specifications of serviceability in cracking for FRP reinforced concrete beams. Compos Struct 182:674–684. https://doi.org/10.1016/j.compstruct.2017.09.016

Jumaa KJ, Ahmed AY, Kareem AA (2019) Shear behaviour of high-strength concrete beams reinforced with basalt FRP (BFRP) bars and stirrups. Constr Build Mater 211:858–872

Junaid MT, Karzad AS, Elbana A, Altoubat S (2022) Experimental study on shear response of GFRP reinforced concrete beams strengthened with externally bonded CFRP sheets. Structure 35:1295–1307

Liang K, Chen L, Shan Z, Su RKL (2023) Experimental and theoretical study on bond behavior of helically wound FRP bars with different rib geometry embedded in ultra-high-performance concrete. Eng Struct 281:115769. https://doi.org/10.1016/j.engstruct.2023.115769

Liew JX, Zhang P, Kai MF, Dai JG (2025) Moisture-induced degradation characteristics of GFRP-concrete interface. Mater Des 255:114183. https://doi.org/10.1016/j.matdes.2025.114183

Luo Y, Liao P, Pan R, Zou J (2024) Effect of bar diameter on bond performance of helically ribbed GFRP bar to UHPC. J Build Eng 91:109577. https://doi.org/10.1016/j.jobe.2024.109577

Lei Z, Luo J, Zhang S, Wang E, Huan H (2024) Mechanical properties of GFRP bars exposed to natural erosion environment: a case study. Case Stud Constr Mater. https://doi.org/10.1016/j.cscm.2024.e03734

Monaco M, Tornabene F, Luciano R, Mele E (2014) Finite element analysis of reinforced concrete beams strengthened with steel truss systems: evaluation of shear transfer mechanisms and nonlinear behaviour. Eng Struct 75:510–525

Moubarak AMR, Ibrahim A, Farouk MA, Elwardany H (2023) Shear performance of glass fiber reinforced concrete beams with diverse embedded GFRP trusses. Case Stud Constr Mater 18:e02195. https://doi.org/10.1016/j.cscm.2023.e02195

Nepomuceno E, Sena-Cruz J, Lourenço L (2025) Assessment of bond durability between GFRP bars and concrete produced with seawater under seawater exposure at different temperatures. Constr Build Mater 469:140479. https://doi.org/10.1016/j.conbuildmat.2025.140479

Noor MA, Jaafar MS, Ali AAA (2012) Shear behaviour of concrete beams reinforced with GFRP bars and stirrups. Mater Struct 45(7):1039–1055

Patil, G. M., & Prakash, S. S. (2024). Effect of macro-synthetic and hybrid fibres on the behaviour of square concrete columns reinforced with GFRP rebars under eccentric compression

Qian W, Lu C, Yuan S, Li H (2025) Experimental and calculation of high-temperature effect on tensile performance of GFRP rebars embedded in concrete. Case Stud Constr Mater 22:e04231. https://doi.org/10.1016/j.cscm.2025.e04231

Qureshi MF, Sheikh SA (2023) Shear behaviour of GFRP strengthened RC beams at elevated temperatures. Constr Build Mater 368:130371. https://doi.org/10.1016/j.conbuildmat.2023.130371

Rahman MM, Cope III CT, Abavisani I, D'Antino T, Focacci F, Carloni C (2025) Comprehensive study of GFRP bar-concrete bond behavior using pull-out and three-point bending tests. J Build Eng 108:112819. https://doi.org/10.1016/j.jobe.2025.112819

Richardson AE (2005) Compressive strengths of concrete with polypropylene fibre additions. Struct Surv 23(2)138–153. https://doi.org/10.1108/02630800510586972

Shabani H, Asadian A, Galal K (2025) Flexural and serviceability behaviour of macro-synthetic fibre-reinforced concrete beams reinforced with GFRP bars. Constr Build Mater 494:143217. https://doi.org/10.1016/j.conbuildmat.2025.143217

Spagnuolo S, Rinaldi Z, Donnini J, Nanni A (2021) Physical, mechanical and durability properties of GFRP bars with modified acrylic resin (modar) matrix. Compos Struct 262:113557. https://doi.org/10.1016/j.compstruct.2021.113557

Solyom S, Balázs GL (2020) Bond of FRP bars with different surface characteristics. Constr Build Mater 264:119839. https://doi.org/10.1016/j.conbuildmat.2020.119839

Starkova O, Aniskevich K, Sevcenko J (2021) Long-term moisture absorption and durability of FRP pultruded rebars. Mater Today Proc 34:36–40. https://doi.org/10.1016/j.matpr.2019.12.154

Sun Y, Jin Z, Zhang X, Pang B (2023) Degradation of GFRP bars in alkaline environments: an experimental and molecular dynamics study. J Build Eng 77. https://doi.org/10.1016/j.jobe.2023.107449

Tesser L, Scotta R (2013) Flexural and shear capacity of composite steel truss and concrete beams with inferior precast concrete base. Eng Struct 55:186–200

Wang H, Belarbi A (2011) Ductility characteristics of fiber-reinforced-concrete beams reinforced with FRP rebars. Constr Build Mater 25:2381–2401. https://doi.org/10.1016/j.conbuildmat.2010.11.040

Wang H, Belarbi A (2013) Flexural durability of FRP bars embedded in fiber-reinforced concrete. Constr Build Mater 44:541–550. https://doi.org/10.1016/j.conbuildmat.2013.02.065

Wang P, Ke L, Leung CKY (2023) Effects of alkaline concentration and saline contents on degradation of tensile properties, microstructure and chemical characterization of GFRP rebars. J Build Eng 69:106222. https://doi.org/10.1016/j.jobe.2023.106222

Wang H, Wu L, Xu X, Lin Z (2024) Experimental study on bond durability of GFRP bar/engineered cementitious composite exposed to freeze-thaw environments. J Build Eng 84:108592. https://doi.org/10.1016/j.jobe.2024.108592

Wei W, Liu F, Xiong Z, Lu Z, Li L (2019) Bond performance between fibre-reinforced polymer bars and concrete under pull-out tests. Constr Build Mater 227:116803. https://doi.org/10.1016/j.conbuildmat.2019.116803

Wei J, Farzadnia N, Khayat KH (2024) Synergistic effect of macro synthetic fiber and shrinkage-reducing admixture on engineering properties of fiber-reinforced super-workable concrete. Constr Build Mater 414:134566. https://doi.org/10.1016/j.conbuildmat.2023.134566

Wiater A, Siwowski T (2020) Serviceability and ultimate behaviour of GFRP reinforced lightweight concrete slabs: experimental test versus code prediction. Compos Struct 239:112020. https://doi.org/10.1016/j.compstruct.2020.112020

Wu W, He X, Alam MS, Han T, Zhang R, Ji K, He J (2023) Comparative research on low cycle fatigue performance of GFRP-RC beams. Case Stud Constr Mater 19:e02317. https://doi.org/10.1016/j.cscm.2023.e02317

Yan F, Lin Z, Yang M (2016) Bond mechanism and bond strength of GFRP bars to concrete: a review. Composites Part B 98:56–69. https://doi.org/10.1016/j.compositesb.2016.04.068

Zhang T, Gao D, Ren M, Xue C (2025) Flexural behavior of hybrid bar reinforced concrete beams and calculation method based on equivalent linear stress-strain relationships. Structure 75:108717. https://doi.org/10.1016/j.istruc.2025.108717

Zheng Z, Feldman D (1995) Synthetic fibre-reinforced concrete. Composites 20:185–210

Zhou J, Chen X, Chen S (2012) Effect of different environments on bond strength of glass fiber-reinforced polymer and steel reinforcing bars. KSCE J Civ Eng 16(6):994–1002. https://doi.org/10.1007/s12205-012-1462-3

Zhu P, Xu J, Qu W, Hao H (2017) Experimental study of fatigue flexural performance of concrete beams reinforced with hybrid GFRP and steel bars. J Compos Constr 21(5):04017036. https://doi.org/10.1061/(ASCE)CC.1943-5614.0000817

Degradation Mechanisms, Durability, Structural Performance, and Design Considerations of Glass Fiber Reinforced Polymer (GFRP) Rebars

C. S. Greeshma, Chandru Pichaimuthu, and R. Kondalraj

Abstract Corrosion of steel reinforcement is one of the primary reasons for the deterioration of reinforced concrete structures, especially in coastal and other aggressive environments. Glass Fiber Reinforced Polymer (GFRP) rebar provides a non-corrosive alternative that improves the durability and long-term performance of concrete structures. Unlike steel, GFRP bars have high tensile strength but low stiffness and do not exhibit yielding before failure. Because of this behaviour, serviceability requirements such as deflection and crack width control govern the design of GFRP-reinforced members. This chapter explains the basic aspects of GFRP reinforcement, such as its composition, how it is manufactured, its mechanical properties, and different surface types. It also looks at how GFRP performs when exposed to moisture, alkaline conditions, marine environments, high temperatures, and fatigue. The bond between GFRP and concrete is explained by focusing on factors such as the surface treatment of the bar, the length embedded inside the concrete, and the level of confinement provided. Structural behaviour of GFRP-reinforced members under flexure and shear are discussed along with the relevant international design guidelines. At present, Indian codes are well developed for the design of steel reinforcement, but guidelines for GFRP as a reinforcement are still limited. Standard codes need to be developed for issues like crack width, bond strength, and long-term durability, especially in coastal/marine regions. The relatively higher cost of GFRP compared to conventional steel can also limit its wider use in construction projects.

Keywords Durability · GFRP rebars · Structural performance · Design · Degradation mechanisms

---

C. S. Greeshma · C. Pichaimuthu (✉) · R. Kondalraj
Department of Civil Engineering, National Institute of Technology Calicut, Calicut, Kerala, India
e-mail: chandru@nitc.ac.in

C. S. Greeshma
e-mail: greeshma_p240079ce@nitc.ac.in

R. Kondalraj
e-mail: kondalrajr@nitc.ac.in

João M. P. Q. Delgado (ed.), *Advanced Materials, Corrosion Science, and Rehabilitation of Concrete Infrastructure*, Building Pathology and Rehabilitation 29,
https://doi.org/10.1007/978-3-032-28851-6_5

1 Introduction

For the past three decades, the use of Glass Fiber Reinforced Polymer (GFRP) bars has been increased in many parts of the world. Today, they are widely viewed as a practical replacement for steel reinforcement in concrete structures. GFRP bars are made of many thin glass fibers placed in the same direction along the length of the bar. These fibers are held together by a resin. The glass fibers carry most of the tensile load, while the resin binds them together, transfers stresses between the fibers, and protects them from environmental damage. Compared to steel, GFRP rebar does not corrode, which greatly enhances the durability and service life of concrete structures. They also have much higher tensile strength about two to three times greater than steel and are much lighter, weighing only about one-fourth as much. In addition, they do not conduct electricity or heat and do not interfere with magnetic or radar signals (Emparanza et al. 2022). Figure 1a shows ribbed GFRP reinforcing bars, and Fig. 1b GFRP bars fabricated as stirrups for shear reinforcement.

The most commonly used glass fibers in GFRP reinforcement are E-glass and S-glass. E-glass is widely used because it is economical, easily available, and provides a good balance of strength, electrical insulation, and chemical resistance. It is suitable for common structural applications where cost and durability are important. S-glass, at the same time, offers higher tensile strength and stiffness and can withstand greater strain before failure. Other types of glass fibers are C-glass which provides better resistance in acidic environments, while alkali-resistant (AR) glass fibers are specially modified, often with zirconium additions to improve durability in the high-pH environment of concrete. AR glass fibers are therefore more suitable for cement-based applications where long-term alkaline resistance is required (Benmoktane et al. 1995). Figure 2 shows that GFRP reinforcement is used in tunnel linings and railway track slabs due to its high corrosion resistance and durability.

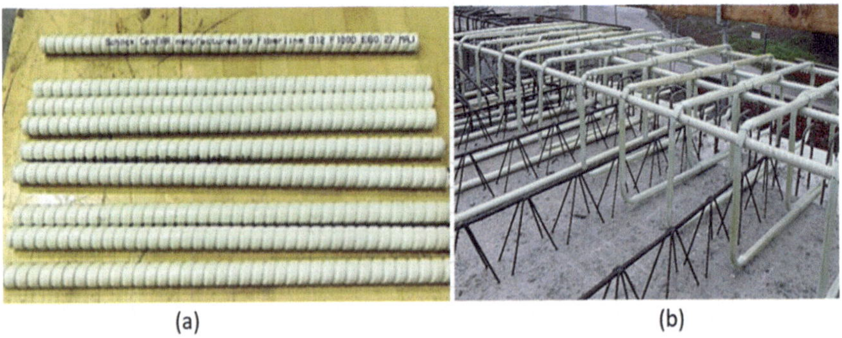

Fig. 1 GFRP rebar

(a) (b)

Fig. 2 Application of GFRP reinforcement in **a** tunnel linings and **b** railway track slabs. Data sourced from dudhi GFRP rebar

1.1 Evolution of FRP Reinforcement

The use of Fiber Reinforced Polymer (FRP) bars in concrete did not begin in civil engineering. FRP materials were first developed in the 1940s and 1950s for aerospace and marine applications because they were lightweight and resistant to corrosion. At that time, the main goal was to reduce the weight of aircraft and ships without losing strength (Hollaway 2010). The use of FRP in concrete structures began much later. In the 1970s and 1980s, many reinforced concrete bridges and marine structures experienced severe corrosion problems. Over time, the steel reinforcement began to rust and caused cracks, surface spalling, and expensive repair work in concrete. Due to these problems, researchers have started their investigation for a new material that does not corrode, and FRP reinforcement have gained their attention here (Benmokrane et al. 2002).

1.1.1 Early Applications (1980s–1990s)

Initially, different types of FRP rebars was used for the structures that has corrosion, especially in marine structures, chemical plants, and bridge decks which are exposed to de-icing. Among these FRP materials, Glass Fiber Reinforced Polymer (GFRP) bars was used widely. One of the main reasons was the relatively lower cost of glass fibers when compared with carbon fibers (ACI Committee 440 2015). However, the first generation FRP materials had several limitations. One of the major concern was the bond between the rebar and concrete was not always consistent, and there was not enough information about how they would perform over a long period in the highly alkaline environment. Engineers were also cautious that about the lower stiffness value of GFRP and they fail in a brittle manner without yielding (Bank 2006). Due to these uncertainties, the use of FRP reinforcement was limited to experimental or trial projects. At that time, it was considered a specialized solution for any specific conditions rather than as a replacement for conventional steel.

1.1.2 Development of Design Guidelines (1990s–2000s)

As research activities increased during the 1990s and early 2000s, there was a significant improvement in the manufacturing and performance of FRP bars. Different surface treatment techniques like sand coating and helical wrapping etc., were introduced to improve the bond performance between FRP bars and concrete. These developments led to the increase of bond behaviour. The results of laboratory and field studies gave a better understanding about its durability and strength retention over time (Benmokrane et al. 2002). A major development occurred when was national design code also validated these results. The American Concrete Institute published the ACI 440 which contains the guidelines for the design of FRP-reinforced concrete members (ACI Committee 440 2015). Thesecode covered the aspects like low stiffness, environmental reduction factors, and serviceability limits etc. Engineers become more confident in using FRP reinforcement in practical structural applications, when design codes and standards was available. Their use and acceptance increased steadily.

1.1.3 Expansion of Applications (2000s–Present)

GFRP bars are now used for various structural applications such as bridges, parking, coastal structures, retaining walls, and water treatment facilities etc. During the last two decades, the use of FRP reinforcement has been extended to many construction projects. Countries like Canada, had adopted GFRP in several infrastructural projects. Their main objective was to reduce long-term maintenance and repair costs in with severe climatic conditions and de-icing salt exposure (Benmokrane et al. 2017a, b). Research is also expanding beyond GFRP. Other types of FRP reinforcement, such as Carbon Fiber Reinforced Polymer (CFRP) and Basalt Fiber Reinforced Polymer (BFRP), is also available for structural applications. These materials have a different mechanical and durability properties, allowing engineers to choose the most suitable reinforcement depending on the structural and environmental Conditions. CFRP bars offer higher stiffness and strength compared to GFRP, but their cost is significantly higher, which limits their widespread use. Basalt Fiber Reinforced Polymer (BFRP) bars have emerged as another option, offering good mechanical properties and improved resistance to chemical attack (Wu et al. 2020). Recent studies focus on long-term durability, creep, fatigue behavior, fire resistance, and bond performance of FRP-reinforced concrete members (Zhang et al. 2021). Improved resin systems and better fiber quality have further enhanced the durability and reliability of modern FRP bars compared to earlier generations.

1.2 Scope and Objectives

Glass fibers are cheaper than carbon or aramid fibers, making GFRP suitable for large-scale structural applications where cost is important. It has high tensile strength, good stiffness, low weight, and is non-conductive, which makes it suitable for aggressive environments such as marine and coastal regions (Elkafrawy et al. 2024). Although reinforced concrete is widely used for its strength and flexibility, steel reinforcement inside it can corrode when chlorides, acids, sulphates, and carbon dioxide penetrate the concrete (Ananthkumar et al. 2020). This corrosion reduces the strength and ductile behaviour of steel, weakens the binding between steelsurface and concrete, and causes cracking due to rust expansion (Hu et al. 2022). GFRP can be considered a suitable option since it does not rust or participate in electrochemical reactions. Because of this, it performs better in aggressive environment and helps to minimize issues such as corrosion-induced cracking and weakening of the concrete bond strength.

The main idea of this chapter is to provide a clear knowledge about GFRP rebar as a non-corrosive material. It tells why corrosion-resistant material is important and how GFRP can overcome the problems related with steel corrosion. The basic material characteristics of GFRP, such as strength, bond behaviour, and durability under different environmental conditions are discussed along with their performance in structural elements. The chapter also discuss various practical challenges related to the use of GFRP, especially in the scenario where indian codes are dominated by the design of steel codes. Finally, the chapter is concluded discussing the sustainability aspects of using GFRP reinforcement.

2 Material Characteristics of GFRP Rebars

The properties of the FRP bars is different from one manufacturer to another. This variation is mainly due to the factors such as the type of fibre, the fibre- resin ratio, and the quality maintained during production (Hajiloo et al. 2018). GFRP bars have a higher tensile strength and better resistance to corrosion. They always shows a satisfactory flexural behaviour. Apart from that GFRP is lightweight and do no rust and maintains good mechanical strength. Because of these properties, GFRP are considered as a suitable alternative to steel, especially in foundation and mass structures where durability is an important requirement (Jabbar and Farid 2018).

2.1 Composition and Manufacturing Process

GFRP rebars are produced by combining continuous glass fibres with an organic resin matrix. During the production, the fibres are first aligned in the longitudinal direction and then impregnated with resin to form a solid composite bar. These bars are commonly available in the diameters ranging from about 8 mm to 20 mm. Smaller diameter have smooth surfaces, whereas larger bars have with ribs or sand coatings on its surface to improve the bonding with concrete. GFRP reinforcement belongs to the family of Fiber Reinforced Polymer (FRP) composite materials (Hassan and Maaddawy 2019).

Composite materials are made up of two parts, the reinforcement phase and the matrix phase. The matrix phase surrounds the reinforcement which helps in the bonding. Depending on the application, the matrix material can be metallic, polymer-based, or sometimes ceramic. The reinforcement, on the other hand, are fibres, particles, or very fine filaments. In fibre-reinforced composites the matrix is usually a polymer resin like epoxy, polyester, phenolic, or vinyl ester. The reinforcing fibres are glass, carbon, or aramid fibres (Kevlar). In general, the fibers act as the primary load-carrying element, providing strength and stiffness. The matrix surrounds and binds the fibers together, transfers load between them, and protects the composite from environmental effects such as moisture, temperature changes, and humidity (Karatas and Gokkaya 2018). The composite material used in GFRP is typically produced by combining about 20% polyurethane resin and 80% glass fiber by mass. The bars are manufactured using the pultrusion process (Gao et al. 2023). Fillers are commonly added to polymer materials to enhance their mechanical properties. The degree of improvement depends on factors such as the stiffness and shape (aspect ratio) of the filler, as well as how well it bonds with the surrounding polymer matrix (Natarajan et al. 2021).

GFRP composites are manufactured by combining glass fibers with a polymer resin using different molding techniques. In general, the process involves placing glass fibers (in the form of mats, woven fabrics, or continuous strands) into a mold. A liquid resin is then added to fully impregnate the fibers. Heat and pressure are applied to cure the resin and form a rigid composite material. After curing, the finished part is removed from the mold. The selection of the method depends on the type of product, required mechanical performance, shape complexity, and production volume. Common manufacturing methods include pultrusion, compression moulding, resin transfer moulding (RTM), vacuum-assisted resin transfer moulding (VARTM), injection moulding, and hand lay-up. Among the available manufacturing methods, pultrusion is the most commonly used process for producing GFRP rebars and other structural sections.

Pultrusion is a continuous and largely automated process used for the manufacture elements with a uniform cross-section, such as rebar. In this process, continuous glass fibres are drawn from spools and arranged in the longitudinal direction. These fibres are then passed through a resin bath, where they are thoroughly impregnated with a polymer resin, like vinyl ester, polyester, or epoxy (Fig. 3).

The resin binds the fibres together and provides protection against environmental exposure. The impregnated fibre bundle is then pulled through a heated steel die. Inside the die, the resin cures under heat and forms a solid composite with the required shape. After curing, the hardened bar is pulled from the die continuously, cooled and cut into the required lengths. In many cases, surface treatments like sand

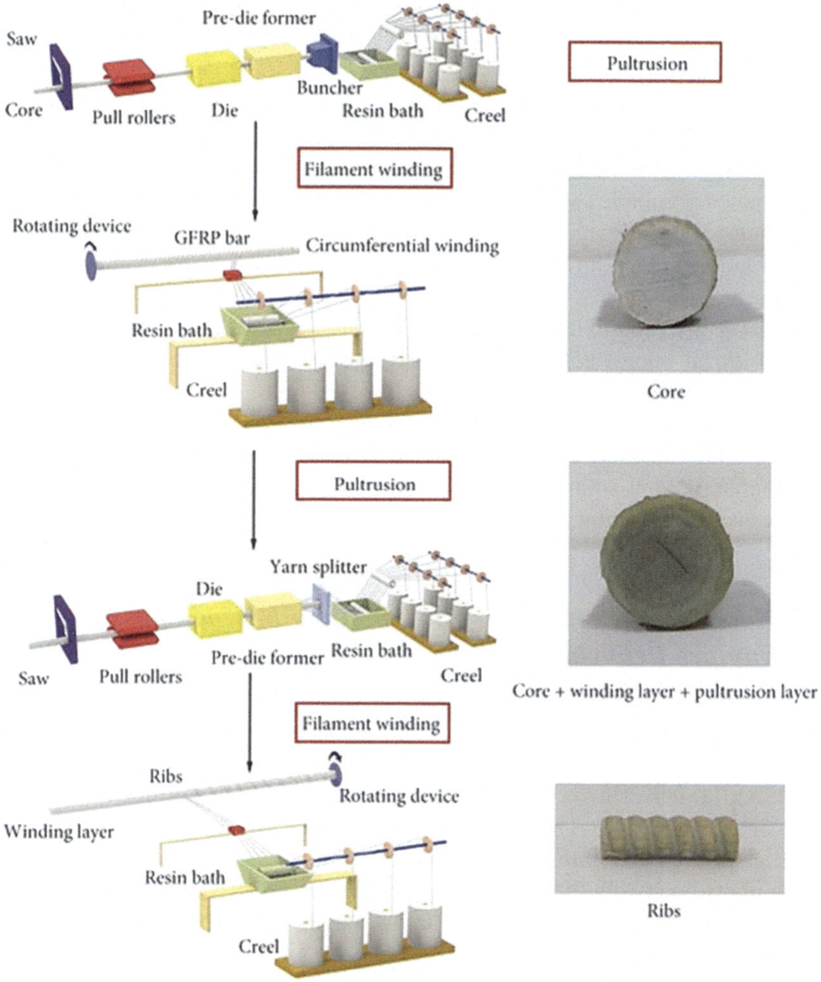

Fig. 3 Schematic illustration of GFRP bar manufacturing process (Liu et al. 2021)

coating or helical wrapping are used to improve the bonding with concrete. Because of these process a bar of uniform size and good strength is produced. Hence pultrusion is widely used method for manufacturing GFRP reinforcement.

(a) In *compression moulding*, glass fibres and resin are placed inside a heated mould. The fibres are arranged in mats or preformed layers before placing in the mould cavity. After positioning, the mould is closed and pressure is applied. Heat and pressure together allow the resin to flow and fill the mould. As the curing takes place, the composite materials gradually hardens and occupy the shape of the mould. This method is generally used for flat or slightly curved components. It produce parts with good surface finish and reasonable dimensional accuracy.

Hence, compression moulding is used for the production of structural and non-structural composite products. However, it is not commonly used for producing GFRP rebars.

(b) In *RTM*, dry glass fiber reinforcement is placed inside a closed mold. A low-viscosity liquid resin is then injected under controlled pressure to impregnate the fibers. After curing, the finished part is removed from the mold. RTM is suitable for complex shapes and medium-volume production. It provides good dimensional accuracy and surface quality. However, it is slower than pultrusion and more suitable for panels or structural components rather than continuous bars.

(c) *VARTM* is similar to RTM but uses vacuum pressure to draw resin into the fiber preform. It improves fiber wetting and reduces void content. This method is commonly used for large components like boat hulls and wind turbine blades.

(d) *Injection moulding* is mainly used for thermoplastic composites. In this process, short glass fibers mixed with molten polymer are injected into a mould. It is suitable for high-volume production of small parts but not for structural rebars because fiber alignment and length are limited (Rajak et al. 2021)

2.2 Mechanical Properties (Tensile Strength, Modulus, Rupture Behavior)

GFRP is an modern composite material that is considered as a potential construction material. It has a high tensile strength than steel and behaves in a linear stress- strain manner until it fails. However, its modulus of elasticity is much lower compared to steel (Ali and Said 2022). The mechanical properties of GFRP mainly depend on the properties of the glass fibres and how wbetween manufacturers due to difference-sell the fibres bond with the surrounding resin matrix. The addition of glass fibers generally increases tensile strength, flexural strength, impact resistance, and elastic modulus. This improvement is primarily due to the higher strength and stiffness of glass fibers and their better bonding with polymer resins. Glass fibers used in GFRP typically exhibit high tensile strength (approximately 2400–4900 MPa depending on fiber type), Young's modulus in the range of about 50–85 GPa, and elongation at break around 4–6%. When glass fibres are combined with polymer resins to produce GFRP composites, the resulting material will have a high strength relative to its weight. It also provides a good stiffness-to-weight ratio, reasonable compressive strength, and better fatigue resistance compared to many traditional construction materials. GFRP also maintains good dimensional stability and mechanical performance under different environmental conditions (Rajak et al. 2021; Sanjay et al. 2015) (Table 1).

The tensile strength of GFRP is around 622 MPa, with an elongation of 4.44%approximately. Its elastic modulus is around 45,979 MPa. The compressive strength is around 311 MPa. The material also shows a flexural strength of about 1,314 MPa, while the flexural modulus is nearly 49,644 MPa (Gao et al. 2023).

Table 1 Mechanical properties of steel and FRP reinforcement (Olczyk et al. 2019; Qureshi 2022)

Property	Steel	GFRP	Unit
Tensile strength	440–550	1000–1500	MPa
Elastic modulus	200–210	50–60	GPa
Unit elongation	10–25	1.5–2.2	%
Deformation	2.18–2.72	30	%
Density	7.85	2.2	g/cm^3
Longitudinal modulus	200	35–60	GPa
Longitudinal tensile strength	450–700	450–1600	MPa
Ultimate tensile strain	5–20	1.2–3.7	%

Figure 4 shows the tensile testing arrangement for the GFRP bar includes a loading head, upper and lower anchorage systems, and a data acquisition unit for monitoring load and deformation.

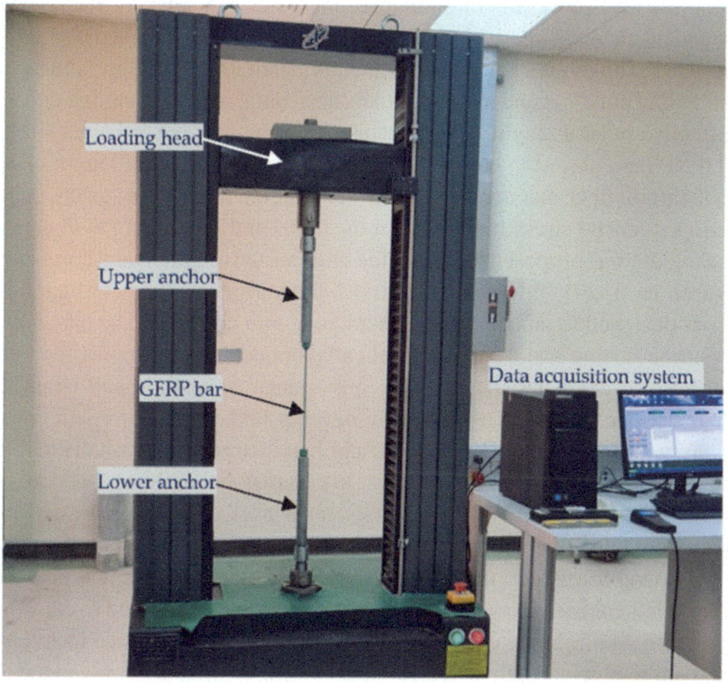

Fig. 4 Tensile test setup of GFRP bar (Alsuhaibani et al. 2024)

2.3 Surface Configurations and Bond Characteristics

Bond characteristics of GFRP bars is mainly controlled by the surface characteristics of the bar rather than the concrete strength. Bond performance is independent of concrete strength, as failure often occurs within the resin layer of the GFRP bar. Low-strength concrete (below 15 MPa), bond strength depends on concrete strength but, for higher strengths (above 30 MPa), failure shifts toward the GFRP surface due to the limited interlayer shear strength of the resin (Baena et al. 2009; Achillides and Pilakoutas 2004; Tepfers 2006). Increasing the strength of concrete does not always improve the bond strength of GFRP bars. Environmental conditions such as high temperature and humidity can reduce bond performance, and long-term aging may also affect bond durability. In addition, bond behaviour is influenced by factors such as the type of FRP used, the resin system, the surface texture of the bars, and the properties of the concrete mix. Studies also indicate that GFRP bars can develop bond strength comparable to or even higher than that of conventional steel reinforcement under appropriate conditions (Czarnecki et al. 2024).

Bond stress transfer occurs through three main mechanisms: chemical adhesion, friction, and mechanical interlocking between the GFRP bar ribs and the surrounding concrete. These forces generate two components: a radial (vertical) pressure acting on the surrounding concrete and a longitudinal (horizontal) component that contributes to the effective bond strength. A study states that the pull-out behavior of 10 mm ribbed GFRP bars placed in concrete, considering an embedment length of 8d and the use of carbon fiber mat anchorage, showed that the anchorage system enhanced the developed tensile stress by 7–21% in medium- and high-strength concrete. An empirical model was proposed to predict the ultimate tensile stress and corresponding slip (Ashrafi et al. 2017). Figure 5 illustrates the various surface configurations of GFRP bars designed to enhance bond interaction with concrete. The bars may have smooth surfaces, fine sand coatings, helically braided fiber bundles, or threaded grooves. Surface characteristics play an important role in the bond behaviour of GFRP bars. Ribbed GFRP bars generally develop higher bond strength compared to helically wrapped and sand-coated helical bars. Ribbed bars usually fail due to concrete crushing and shearing of the ribs. In contrast, helically wrapped and sand-coated bars often fail because the fibre spirals, resin layer, or sand coating detach from the bar surface, while the surrounding concrete remains mostly intact. Sand-coated helical bars demonstrated better bond performance than conventional helical bars. Sand-coated helical bars showed better bond performance compared to conventional helical bars. In addition, the use of stirrups changed the failure mode from concrete splitting to pullout, which helped increase the bond strength and improved the overall bond behaviour (Saad and Polak 2025).

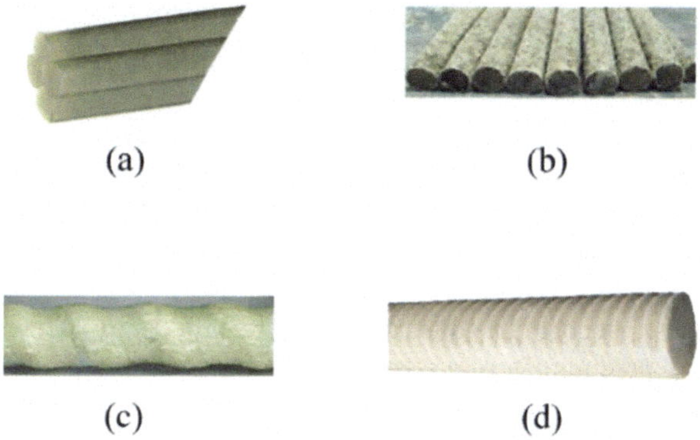

Fig. 5 Surface Configurations of GFRP Bars **a** Smooth-surfaced GFRP bar **b** Fine sand-coated GFRP bar **c** GFRP bar with helically braided fiber bundles **d** GFRP bar with threaded grooves (Nguyen et al. 2025)

2.4 Comparison with Steel Reinforcement

GFRP reinforcement has attracted attention in structural applications as a corrosion-resistant alternative to steel, especially in aggressive exposure conditions. However, its mechanical behaviour is different from steel in terms of strength, stiffness, and deformation. For this reason, a clear comparison between the two materials is important for proper structural design.

- Tensile Strength: GFRP bars demonstrated approximately 13% higher tensile strength than steel bars. In addition, their yield strain was about 58% higher than that of steel, indicating greater deformation capacity before failure.
- Bending Strength of Bars: The flexural strength of GFRP bars is about 72% of that of steel bars. However, GFRP bars show nearly 20% higher strain at failure compared to the yield strain of steel, which indicates greater deformation before breaking.
- Flexural Strength of RC Beams**:** Concrete beams reinforced with GFRP generally show lower flexural strength than steel-reinforced beams. Beams with smooth GFRP bars reached about 71–75% of the strength of steel RC beams, while sand-coated GFRP beams performed slightly better, reaching around 77–82%.
- Flexural Modulus (Stiffness) of RC Beams**:** GFRP-reinforced beams have lower stiffness compared to steel RC beams. The flexural modulus was reduced by about 66% for smooth bars and around 33% for sand-coated bars when compared with steel-reinforced beams.
- Flexural Strain Capacity of RC Beams**:** At failure, GFRP-reinforced beams showed higher flexural strain than steel RC beams, indicating greater deformation before rupture.

- The increase in strain was about 44% for smooth GFRP bars and about 14% for sand-coated GFRP bars, indicating greater deformation capacity before failure (Jabbar and Farid 2018).
- Service Stress Limits: Steel bars are typically limited to about 60% of yield stress at service load, whereas GFRP bars are limited to 20–30% of their tensile strength depending on design codes. Steel reinforcement is generally limited to about 1200 με under service conditions, while GFRP bars are permitted up to about 2000 με, though at lower stress levels.
- Design Control: Steel RC design is primarily strength-controlled, whereas GFRP RC design is predominantly governed by serviceability requirements such as deflection and crack control. To control crack width (generally below 0.7 mm), the ISIS design manual limits GFRP strain to 2000 με under service conditions (Nkurunziza et al. 2005)

3 Durability Behavior of GFRP Rebars

GFRP bars can be affected by certain environmental conditions, especially alkaline environments, moisture exposure, and high temperatures. These conditions may lead to a reduction in tensile strength, ultimate strain, and elastic modulus.

3.1 Moisture and Alkali R5esistance

Most GFRP bars are made with E-glass fibers, which are vulnerable to alkaline attack, therefore, durability largely depends on the resin matrix which limits the penetration of harmful ions. Moisture absorption is influenced by the type of polymer matrix and, can lead to plasticization, strength reduction, and a decrease in glass transition temperature due to molecular-level diffusion within the matrix (Antino et al. 2018; Benmokrane et al. 2017). Moisture absorption primarily affects stiffness of GFRP and is largely controlled by the resin matrix. Flexural strength decreases more slowly, since it is primarily governed by the fibers, unless the fibers themselves become damaged. Long-term water exposure can cause ions to leach from the glass fibers, leading to surface damage and early failure. As a result, the retention ratio gradually declines over time, indicating continuous deterioration of GFRP properties under hydrothermal exposure.

Moisture absorption in epoxy-based composites produces a plasticizing effect, leading to a reduction in the glass transition temperature (Tg) of the matrix. This effect is generally reversible once the moisture is removed. This type of degradation can create small internal voids within the polymer matrix. It may also cause expansion of the polymer chains and lead to the formation of micro cracks. The amount of moisture absorbed depends on several factors, such as temperature, fibre volume content, fibre orientation, type of reinforcement, exposed surface area, material diffusivity, and whether any surface protection is provided etc (Sateesh et al. 2015).

When GFRP rebar is submerged in water or alkaline solutions, the polymer matrix absorbs moisture. This can weaken the bond between the fibres and the matrix, leading to delamination and the development of cracks within the composite structure. These defects act as pathways for further ingress of water and aggressive chemicals, accelerating degradation and increasing pore size and number over time. The resulting damage weakens stress transfer between fibers and allows harmful substances to reach the fiber surface, causing leaching and etching of the glass fibers. This process deteriorates the fiber–matrix interface and ultimately reduces the mechanical performance of the GFRP rebar. SEM observations confirmed that degradation was much more severe in alkaline conditions, while exposure to tap water had only a minor effect on the material performance. SEM analysis revealed that microstructural degradation of GFRP rebar was associated with an increase in pore formation, leading to matrix delamination and cracking. This process also reduced the effective cross-sectional area of the glass fibers, ultimately resulting in a decline in the mechanical performance of the GFRP rebar (Won et al. 2008). Figure 6 presents the surface degradation of GFRP bars after prolonged exposure to an alkaline solution at 60 °C. The SEM images at 180, 240, and 300 days reveal progressive matrix deterioration, fiber exposure, and micro cracking.

3.2 Chloride and Marine Exposure

Exposure to harsh environments like alkali and salts affect the performance of FRP bars by softening the resin matrix, damage the fibres, and weaken the bond between the fibre and the resin. In marine or coastal environments, FRP bars are exposed to seawater, higher temperatures, the alkaline pore solution of concrete, and sustained loading. The combined action of these increases the possibility of internal damage and can reduce the load-carrying capacity of the composite material. When FRP bars remain in seawater for a long periods, water molecules and ions such as chloride and sodium slowly enter the matrix. This happens through diffusion and capillary movement. As these substances penetrate, the bond between the fibre and resin interface weakens gradually. Fibres start to damage. Also the resin swell slightly, and small cracks develop inside the composite. With the prolonged exposure, more and more ions accumulate at the interface. This leads to the formation of a hydrated layer that disturbs the original chemical composition and reduces its strength. Higher temperatures increase moisture absorption and accelerate cracking. The alkaline nature of concrete can further increase the rate of deterioration. Hydroxyl ions in the pore solution react with the resin and attack the silica present in glass fibres, which again weakens the material. Using dense concrete with a lower permeability and lower alkalinity can overcome this issue. This can also slow down the penetration of aggressive ions and the improve long-term durability. The combined effect of sustained stress and environmental condition is also important. When the applied

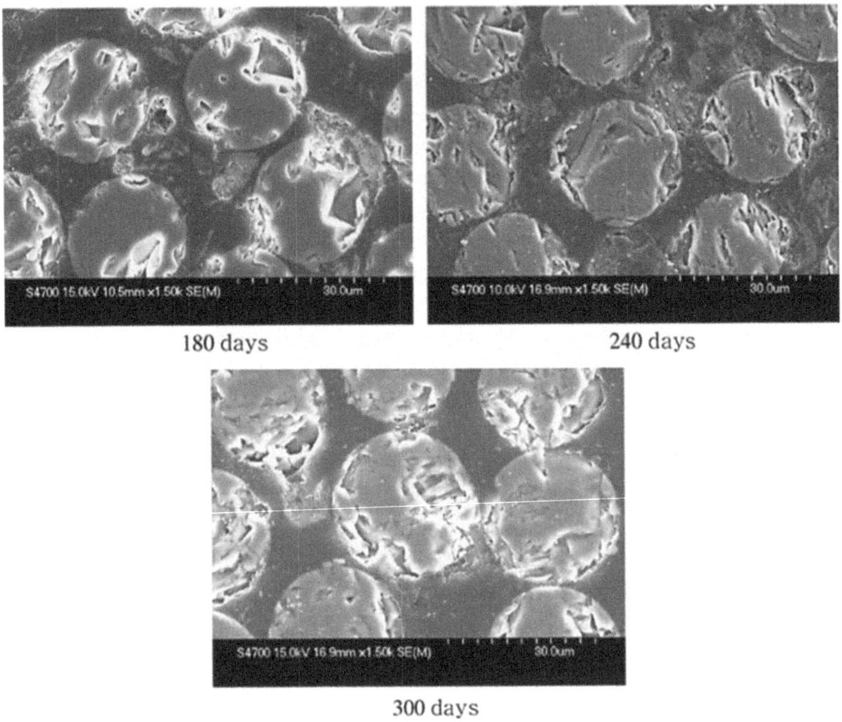

Fig. 6 Surface degradation of GFRP after exposure to alkaline solution at 60 °C (Won et al. 2008)

stress is low, around 0–20% of the ultimate strength, degradation occurs due to slow diffusion of aggressive ions. At moderate stress levels, roughly between 20 and 40%, micro cracks start appearing in the resin and fibre–matrix interface, which increases the rate of damage. If the stress level again becomes higher, about 40% of the ultimate strength, the material tends to behave more brittle and failure may occur suddenly (Wen and Cao 2023).

Marine structures are subjected to repeated wet- dry cycles because of tidal action. Therefore, it is important to study how GFRP bars perform under such exposure conditions. Earlier studies have reported that bare GFRP bars were able to retain nearly 90% of their tensile strength even after long-term exposure to seawater wet–dry cycles at elevated temperatures, which suggests reasonably good durability. However, most of these investigations did not take the effect of concrete cover into account. In actual structures, the internal alkaline environment of concrete interacts with external moisture and humidity. Because of this interaction, the presence of concrete cover may influence the degradation behaviour of GFRP bars over time. Seawater–sea sand concrete (SWSSC) reinforced with GFRP bars offers a sustainable solution for marine construction. SWSSC worked effectively with GFRP bars, demonstrating good overall durability in chloride environments. Although sustained loading and exposure led to some prestress loss, the ultimate flexural strength of over-reinforced

Fig. 7 GFRP rebars exposed to marine conditions (Chen et al. 2021)

beams did not reduce significantly, as failure was governed mainly by the concrete (Wang et al. 2023).

The deterioration of GFRP in marine environments is mainly attributed to moisture ingress and hydrolysis of the outer layer. This process weakens the bond between the fiber and matrix, gradually reducing the thickness of the protective FRP layer. As a result, more fibers become exposed and the protection provided to the inner steel bar decreases. Under chloride exposure, the tensile response of GFRP follows three stages: an initial elastic stage, a pseudo-strain hardening stage, and a final rapid fracture stage. This behavior remains similar to that of unaged specimens, indicating that seawater exposure does not alter the fundamental failure mode. While the equivalent elastic modulus remains largely unaffected, the pseudo-strain hardening modulus decreases with prolonged exposure. The tensile strength retention declines during the early stages of aging but stabilizes over time. This suggests that chloride attack mainly affects the outer surface of the FRP jacket, whereas the inner layers exhibit relatively good resistance to further degradation (Cai et al. 2023). Figure 7 presents the condition of GFRP rebars after exposure to marine environments. In comparison with the unexposed bar, specimens subjected to tidal conditions for 180 and 270 days show noticeable surface discoloration and deterioration.

3.3 *Elevated Temperature and Fire Performance*

Glass fibres are able to maintain their strength even at relatively high temperatures, as the softening point of E-glass is around 850 °C. However, the polymer matrix in GFRP is much more sensitive to heat. At elevated temperatures, the matrix undergoes glass transition and eventual decomposition, leading to significant changes in its thermal and mechanical properties. Since GFRP is a composite material made of glass fibres embedded in a polymer matrix, any deterioration of the matrix directly affects the overall performance of the material. Moreover, the changes that occur at high temperatures are time-dependent, as they are governed by chemical and physical processes such as glass transition and thermal degradation (Dwight 2000; Ashby et al. 1997; Correia et al. 2015).

GFRP specimens exposed to elevated temperatures exhibit progressive physical and structural changes. As shown in Fig. 8, specimens subjected to temperatures from 20 to 300 °C display increasing discoloration, surface cracking, and thermal degradation. During fire exposure, issues such as smoke release, toxic gases, or resin dripping are important safety concerns. However, the primary consideration for pultruded GFRP profiles is their ability to maintain structural integrity and load-carrying capacity. Fire resistance largely depends on the type of loading (tension or compression), the cross-sectional configuration (open or closed sections), wall thickness, and the number of surfaces exposed to fire. Based on these factors, fire resistance can range from only a few minutes to nearly one hour. In many cases, fire resistance is governed by the glass transition temperature of the resin, particularly in compression zones. When the resin reaches its glass transition temperature often around 100 °C the material rapidly loses stiffness and load-bearing capacity. The use of phenolic resins or fire retardants does not significantly improve performance in compression-critical regions. Therefore, such areas require adequate fire protection, unless the structural system is designed with redundancy, allowing localized failure without leading to overall collapse (Mouritz et al. 2006; Correia et al. 2015).

Increasing the glass fibre content in a composite can improve its fire resistance, as glass fibres themselves do not burn. When the proportion of glass fibres is higher, there is less organic material available to produce flammable gases, which delays ignition. On the other hand, some types of reinforcement that contain a higher amount of organic binding agents, may ignite more quickly. Increasing the thickness of the laminate can delay ignition up to a certain limit, mainly because thicker sections have greater thermal capacity and take longer to heat up (Brown and Mathys 1997; Scudamore 1994; Mouritz et al. 2006; Correia et al. 2015).

Fig. 8 GFRP specimens after exposure to elevated temperatures (20–300 °C), showing progressive surface degradation, discoloration, and thermal damage (Aybar et al. 2025)

The fire performance of FRP can be improved in two main ways: first, by adding flame-retardant materials into the composite and second by applying protective layers such as fire-resistant coatings or boards (Correia et al. 2015). Common flame-retardant additives used in FRP include silica, calcium carbonate, talc, aluminium trihydroxide (ATH), magnesium hydroxide, antimony oxide, and zinc borates. These materials help reduce flammability and enhance fire resistance (Mouritz et al. 2006). Intumescent coatings are generally effective in slowing or controlling fire spread, but their performance depends on the type and thickness of the coating. Epoxy-based intumescent coatings, however, may contribute to fire growth in the early stage because they are partly combustible and can release heat before a protective char layer is fully formed (Ohlemiller and Shields 1999).

3.4 *Creep Rupture and Fatigue Behavior*

Pultruded GFRP members are frequently exposed to harsh outdoor conditions and sustained loads over long periods. Under such conditions, they may experience excessive creep deformation, which in severe cases can lead to structural instability or collapse. For this reason, creep behavior is a major concern in long-term applications. Creep generally refers to time-dependent deformation that occurs under constant stress, typically at elevated temperatures and at stress levels below the material's yield strength. In polymers, however, this deformation is primarily viscoelastic in nature, meaning the material shows both elastic and time-dependent characteristics. In FRP materials, creep is particularly important because prolonged exposure to sustained stresses can reduce stiffness, compromise durability, and in some cases lead to permanent deformation or buckling of structural elements (Alhayek et al. 2022).

Composite materials are heterogeneous and anisotropic, as their properties vary in different directions and locations. As a result, damage in composites tends to spread over a wider area rather than remaining localized, as commonly seen in metals. In metals, failure usually begins with the growth of a single crack whereas composite materials experience multiple damage mechanisms, such as fiber breakage, matrix cracking, delamination, fiber–matrix debonding, transverse cracking, or a combination of these modes (Gaurav and Singh 2018). Large-scale fatigue testing has demonstrated that GFRP-reinforced beams and slabs can endure millions of load cycles with minimal loss of stiffness, indicating their suitability for railway-type repeated loading conditions. Studies on bond performance under cyclic loading have shown that the GFRP–concrete interface remains stable after extensive load repetitions, ensuring reliable stress transfer at service levels. Fatigue durability is strongly influenced by the fibre volume and overall composite configuration. Full-scale comparative studies between steel- and GFRP-reinforced slab tracks have shown that GFRP reinforcement can extend fatigue life by over 30% under identical loading conditions, indicating superior durability and improved resistance to repeated loading in railway structures (Lee 2025).

Laminated composites experience creep even at room temperature because the polymer matrix behaves in a viscoelastic manner. When the temperature remains below the glass transition temperature (Tg) of the resin, creep mainly occurs due to time-dependent deformation of the matrix. This can lead to resin micro cracking (crazing), fiber straightening under tension, fiber buckling under compression, de-bonding at the fiber- matrix interface, fiber fracture, and eventually overall failure. Conventional creep under sustained load, laminated composites may also undergo cyclic creep when subjected to fatigue loading with a non-zero mean stress (Vieille et al. 2014; Wong et al. 2014; Movahedi-Rad et al. 2019).

3.5 Long-Term Degradation Mechanisms

Among the various exposure conditions studied, moist and alkaline environments are considered the most harmful to GFRP composites. In moist conditions, hydroxide ions (OH^-) attack the polymer matrix, breaking down its long molecular chains through hydrolysis. This process allows more water molecules and hydroxide ions to penetrate the material, further accelerating degradation. When the polymer chains start to break down, the weaker secondary bonds between them are also affected. This creates small internal voids and increases the free space within the material. As a result, the matrix becomes plasticized, and its mechanical and physical properties gradually decrease.

Water exposure can also affect the glass fibres. Through a leaching process, alkali components are slowly removed from the glass structure. The hydroxide ions released during this reaction increase the alkalinity of the surrounding solution. When the pH exceeds above 9, the silica network in the glass fibres begins to dissolve, damage the fibres and reduce the overall performance of the composite (Fergani et al. 2018).

In GFRP-reinforced concrete, bond degradation primarily occurs due to two reasons. First, deterioration of the surrounding concrete which weakens the bond layer holding the GFRP bar. Second, damage to the surface ribs of the GFRP that reduces mechanical interlocking, and lead to bond failure. As per the scanning electron microscope (SEM) images it is evident that in ordinary concrete, micro cracks develop due to shrinkage and creep, which further reduce bond strength. From a chemical perspective, degradation begins when the resin on the outer surface reacts with water and hydroxide (OH^-) ions, leading to hydrolysis of the polymer chains. The Si–O–Si bonds in the glass fibres also break down through similar reactions. In a highly alkaline environment, where the concentration of OH^- ions is higher, these reactions occur more rapidly. Therefore, alkaline conditions cause more severe deterioration of GFRP bars compared to seawater or saline environments (Wu et al. 2014).

4 Bond Behaviour and Interaction with Concrete

Bond failure of GFRP bars mainly depends on factors such as concrete cover, confinement, embedment length, bar diameter, surface condition, and concrete strength. If these parameters are properly designed and detailed, pullout failure can be achieved. Pull out is generally preferred because it is more gradual and controlled compared to sudden splitting of the surrounding concrete.

4.1 Bond Mechanism and Failure Modes

The bond behaviour of GFRP bars in concrete is an important factor that influences their performance in corrosion-resistant reinforced concrete structures. Unlike steel, GFRP is anisotropic, non-homogeneous, and exhibits linear elastic behaviour up to failure. Because of these material characteristics, the mechanism of force transfers between GFRP bars and surrounding concrete differs from that of conventional steel reinforcement. Several factors influence the bond performance of GFRP bars, including the compressive strength of concrete, the thickness of concrete cover, and the level of confinement provided by transverse reinforcement (Brown et al. 1993; Pecce et al. 2001; Yan et al. 2016).

The bond between reinforcing bars and concrete generally consists of three components: (a) chemical adhesion at the interface, (b) frictional resistance against slip, and (c) mechanical interlocking caused by surface irregularities of the bar. GFRP bars exhibit bond behaviour that differs from steel reinforcement, mainly due to differences in material properties and surface characteristics. Unlike steel, which shows a clear plastic stage after yielding, GFRP behaves in a linear elastic manner until failure. GFRP bars are manufactured with various surface treatments such as ribbed, sand-coated, or helically wrapped finishes. Despite these treatments, the mechanical interlocking effect of GFRP bars is generally lower than that of conventional ribbed steel bars. Many studies suggest that mechanical interlocking provided by the surface texture plays a major role in the bond strength of GFRP bars, while chemical adhesion and friction contribute to a lesser extent. However, as per the researchers chemical adhesion, control the bond during the early stages of pull out, while mechanical interlock and friction act as secondary mechanisms (Cosenza et al. 1997; Chaallal et al. 1993; Yan et al. 2016).

Pull-out failure and splitting failure are the dominant failure modes of GFRP bars in concrete are, together account for more than 80% of reported cases. Pull-out failure occurs, when adequate concrete cover and embedment length are not provided. Splitting failure happens when the concrete cover or confinement is insufficient, causing cracks to form in the surrounding concrete, before the ultimate bond strength is reached (Benmokrane et al. 1996; Makitani et al. 1993; Yan et al. 2016) (Table 2).

Other less common failure modes include anchorage failure, tensile rupture of the bar, and peeling of the outer resin layer. Anchorage failure is more likely in

Table 2 Failure modes of GFRP-reinforced concrete members (El-Nemr et al. 2018; ACI 440)

Failure mode	Governing condition	Description	Observed In
Concrete crushing	$\rho f > \rho fb$	Compression failure	Majority of tests
Bar rupture	$\rho f < \rho fb$	Sudden tensile rupture	Under-reinforced sections
Bond splitting	Low cover/poor confinement	Concrete cracking	Bond tests

higher-strength concrete. Tensile fracture of the bar occurs when the bond strength exceeds the tensile capacity of the GFRP bar. Peeling of the resin layer is usually seen in specimens with higher concrete strength, where stress concentrations at the interface damage the outer surface of the bar.

Concrete strength has a clear influence on the type of failure. At normal strength levels (about 30–50 MPa), pullout and splitting failures are more common. At higher concrete strengths, failure tends to be more brittle, and bond slip reduces, which may result in sudden splitting or surface damage. Concrete cover is also important. When the cover is insufficient, splitting failure becomes more likely.

As the cover increases (typically $\geq$ 2–4 times the bar diameter), pullout failure becomes dominant due to improved confinement (Ehsani et al. 1993; Alves et al. 2010; Yan et al. 2016). Figure 9a presents the experimental setup used for the pull-out test of GFRP bars, while Fig. 9b provides a schematic representation of the loading frame, hinge mechanism, concrete block, and LVDTs used to measure slip and displacement.

4.2 *Factors Influencing Bond Strength*

Bond strength describes the interaction and load transfer mechanism between two materials, such as GFRP reinforcement and the surrounding concrete, and how they work together under applied loads. The quality of this bond is essential for ensuring effective stress transfer, structural stability, and overall integrity of reinforced concrete systems (Devaraj et al. 2023). Figure 10 compares the bond pefor-mance of steel and GFRP bars. As shown in Fig. 10a, steel bars exhibit higher bond strength than GFRP bars, Fig. 10b illustrates the bond stress-slip response, where steel reinforcement shows a higher peak bond stress with a sharper decline, while GFRP bars demonstrate a comparatively lower but more gradual bond behaviour. The bond strength of GFRP bars in concrete is affected by several key parameters.

- Concrete compressive strength: Bond strength increases approximately in proportion to $\sqrt{f'c}$. However, when concrete strength exceeds about 55 MPa, the normalized bond strength ($tb/\sqrt{f'c}$) may decrease due to brittle splitting behaviour (Azizinamini et al. 1993; Zuo et al. 2000; Yan et al. 2016).

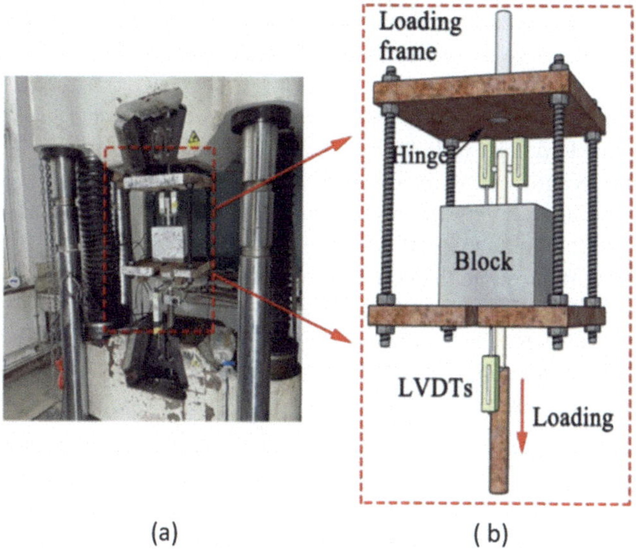

Fig. 9 Experimental setup showing **a** pull out test of GFRP **b** Schematic diagram of loading frame, hinge, block, and LVDTs (Hu et al. 2023)

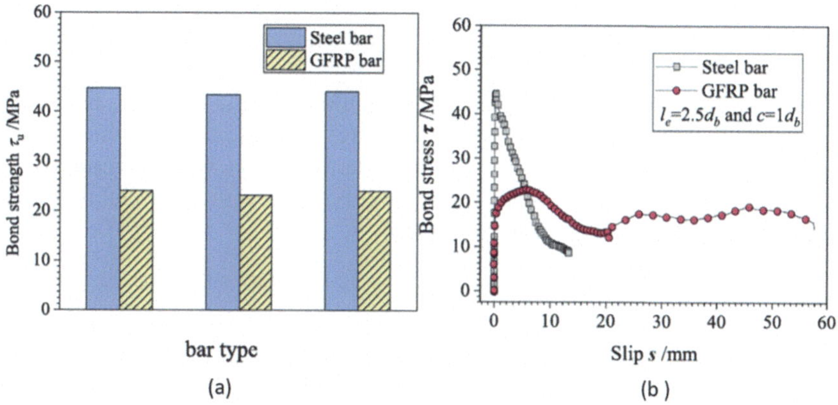

Fig. 10 Comparison of bond performance between steel and GFRP bars: **a** bond strength for different bar types, and **b** bond stress–slip response showing the variation in bond behaviour with slip for steel and GFRP reinforcement (Hu et al. 2023)

- Surface Morphology: Surface morphology strongly influences the bond behaviour of GFRP bars. The type of surface finish on GFRP bars has a strong influence on how they bond with concrete. For sand-coated bars, the bond strength is due to friction between the rough surface and the surrounding concrete. In contrast, ribbed and helically wrapped bars develop bond strength primarily through mechanical interlocking between the surface projections and the concrete (Devaraj et al. 2023).

- Concrete cover: Increasing concrete cover improves confinement and shifts failure from splitting to pullout (Aly 2009; Yan et al. 2016).
- Embedment length: As embedment length increases, the average bond stress decreases because bond stresses are not uniformly distributed along the bar (Robert and Benmokrane 2010; Makitani et al. 1993; Yan et al. 2016).
- Bar diameter: Bond strength decreases with increasing bar diameter due to higher probability of interface defects and stress concentration.
- Transverse reinforcement confinement: Confinement from stirrups can delay splitting and enhance ductility, although its influence on average bond stress for FRP bars remains debated.
- Casting position: Top-cast bars show reduced bond strength compared to bottom-cast bars. uHu HH

4.3 Anchorage and Development Length

Development length refers to the minimum length of reinforcement that must be embedded within the concrete so that the bar can safely develop its full design strength at a critical section without bond failure (Arefian and Mostofinejad 2025). Embedment length also affects failure mode. An embedment length of about 5–6 times the bar diameter generally promotes pullout failure. However, increasing embedment length reduces the average bond stress because bond stresses are not uniformly distributed along the bar. Bar diameter influences bond performance as well. Larger diameters are associated with lower bond strength and a slightly higher tendency for splitting failure (Yan et al. 2016).

4.4 Influence of Surface Treatment

Surface condition significantly affects failure mode. Ribbed, helically wrapped, and sand-coated bars improve mechanical interlocking, often leading to pullout failure. However, stronger surface interlocking may also increase hoop stresses in concrete and induce splitting if confinement is insufficient (Yan et al. 2016). As GFRP bars are manufactured with different surface configurations, such as sand-coated and ribbed profiles, it is important to evaluate how surface morphology influences bond performance in combination with other governing parameters. When ribbed GFRP bars are used, bond resistance is provided by both chemical adhesion and mechanical interlocking between the ribs and the surrounding concrete. After the chemical bond breaks, the ribs continue to resist pull-out through mechanical interlock, which governs the bond strength. This increased pull-out resistance generates tensile stresses in the surrounding concrete, and once the tensile capacity of the concrete is exceeded, splitting failure occurs.

For sand-coated GFRP bars, bond strength mainly depends on the interaction between the epoxy-bonded sand layer and the surrounding concrete and how well the sand coating is bonded to the core of the bar. Failure starts when the bond between the sand coating and the concrete start to weaken. After that, the coating may separate from the bar core, leading to noticeable slip of the bar. Ribbed GFRP bars demonstrated the highest bond performance, achieving a bond energy of 89.4 Nmm and an average bond strength of 11.9 MPa. Failure mode analysis further highlighted the significant role of surface characteristics. All ribbed GFRP bars failed by concrete splitting, whereas approximately 85% of sand-coated bars failed due to bar slippage (Devaraj et al. 2023).

5 Structural Performance of GFRP Reinforced Concrete Members

5.1 Flexural Behavior

FRP materials behave in a linear elastic manner under tension and fail suddenly without yielding. Unlike steel, which undergoes plastic deformation before failure, FRP does not provide warning through large inelastic strains. In steel-reinforced concrete members, ductility is mainly achieved when the steel yields, allowing the structure to absorb and dissipate significant energy while the concrete reaches its full compressive strain capacity. In contrast, FRP-reinforced members lack this yielding mechanism, which can result in reduced structural ductility and a exhibit brittle type of failure (Li and Wang 2002). Glass fibres can reach tensile strengths of up to 4.9 GPa and an elastic modulus of about 86.9 GPa. These values indicate that glass fibres have significantly higher tensile strength than conventional steel reinforcement. However, they possess a much lower density and a lower elastic modulus compared to steel (Tran et al. 2025).

Several studies have investigated the flexural performance of GFRP-reinforced beams. In one study when the compressive strength of concrete increased from 24.5 to 74.4 MPa, the flexural capacity of GFRP-reinforced beams increased by 84.3%, while the crack width reduced by 80.0%. Another investigation on ordinary and high-strength GFRP-reinforced beams showed that beams with higher concrete strength exhibited a greater ductility index (Adam et al. 2015; Abdelkarim et al. 2019). For ultra-high performance concrete (UHPC), GFRP beams have exhibit substantially higher bending capacity compared to steel-reinforced beams up to about 70% higher in some cases although this improvement is accompanied by increased mid-span deflection and reduced flexural stiffness (Li et al. 2024). The stress–strain curves of Fig. 11 shows GFRP bars exhibit linear elastic behavior up to failure without yielding,

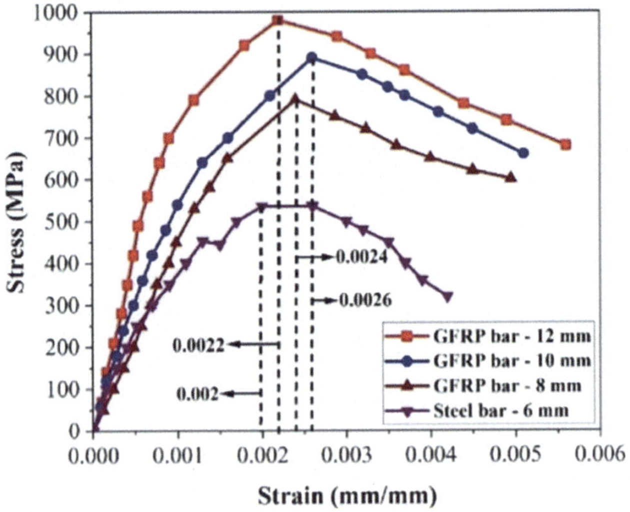

Fig. 11 Stress- strain behaviour of steel and GFRP bar (Sasikumar and Manju 2024)

while steel demonstrates yielding and plastic deformation. GFRP bars have higher tensile strength but lower ductility compared to steel.

5.2 *Shear Behaviour*

In FRP-reinforced members, higher reinforcement ratios are often provided to control deflection. As a result, flexural tension failure is usually prevented, and the member becomes more prone to shear failure. When beams are subjected to both bending and shear, failure can occur either due to rupture of the shear reinforcement (or yielding in the case of steel) or due to crushing of the diagonal compression zone in concrete. Failure caused by rupture of FRP shear reinforcement is highly brittle, since FRP bars have limited capacity to resist shear stresses acting across them. Failure governed by compression of the concrete can allow some deformation before collapse, similar to flexural compression failure. Therefore, an over-reinforced design that leads to concrete compression failure is generally preferred for FRP-reinforced members. However, the ductility achieved in such beams depends on the level of confinement provided in the compression zone (Li and Wang 2002).

For GFRP-reinforced concrete beams designed to fail by concrete crushing (compression-controlled failure), the flexural capacity can be estimated using the ultimate strength approach applicable to over-reinforced sections. In many cases, service load deflection governs the design of FRP-RC members (Alsayed 1998). The shear resistance in GFRP beams is less reliable and more brittle than in steel

RC, especially because of reduced dowel action, uneven stress distribution in stirrups, and weakness at bent stirrup regions. The bent portions of GFRP stirrups are relatively weaker than the straight segments and are more susceptible to localized failure at load levels lower than those expected based on the tensile strength of the straight bar sections. The presence of stirrups significantly improves shear performance, increasing the ultimate shear capacity by about 2.76 times and the vertical displacement at failure by about 3.58 times compared to beams without stirrups. Shear resistance of GFRP-RC beams resulted the from the combined action of dowel effect, aggregate interlock, uncracked compression zone, residual tensile stresses of concrete, and stirrups (Resende et al. 2025).

5.3 *Deflection and Serviceability*

Serviceability considerations play a dominant role in the behaviour of GFRP-reinforced concrete beams. In the initial stages of loading, cracks will occur within the constant moment region. As loading progress, additional flexural cracks developed at midspan, following the formation of vertical cracks within the shear span. However, the crack development and ultimate behaviour differed significantly between steel-reinforced and GFRP-reinforced beams. GFRP-reinforced beams fail primarily due to bond-slip mechanisms rather than concrete crushing. Vertical cracks within the shear span may became critical and extend toward the loading point, accompanied by splitting cracks parallel to the reinforcement level. Hence the slip of the GFRP bar from the surrounding concrete governed failure, resulting in ultimate loads lower than the corresponding design loads. The load–deflection behaviour also reflects the influence of reinforcement stiffness on service performance. Prior to cracking, the stiffness of all beams was governed by gross section properties. After cracking, however, the stiffness depended largely on the longitudinal reinforcement (Saikia et al. 2007).

Increasing the FRP reinforcement ratio has a stronger effect on the service moment than on the ultimate moment capacity. Similarly, reducing the spacing between bars increases the service moment capacity, while the ultimate resistance moment is primarily enhanced by increasing the concrete strength (Abdelkarim et al. 2019). The deflection behaviour of GFRP-reinforced concrete beams has been linked to a rapid reduction in stiffness after cracking, mainly due to the low elastic modulus of GFRP bars. To address the limited ductility associated with FRP reinforcement, researchers have investigated hybrid systems in which a portion of the steel reinforcement is replaced with FRP bars (Maranan et al. 2019; Shabani et al. 2025).

5.4 Crack Width Control

FRP-reinforced concrete beams generally develop wider cracks than steel-reinforced beams. This mainly happens because FRP bars have a lower modulus of elasticity. As a result, design is often controlled by serviceability limits, particularly crack width and deflection. The service moment as well as the ultimate moment capacity depend on factors such as bar diameter, bar spacing, and concrete strength. Increasing the concrete strength and the reinforcement ratio can help reduce crack width, as it leads to the formation of more closely spaced cracks along the beam. Bar diameter does not significantly affect the overall deflection. However, larger bar diameters tend to produce wider cracks, and this effect is more noticeable in normal-strength concrete (NSC) beams (Abdelkarim et al. 2019) (Table 3).

High-strength concrete (HSC) beams demonstrated greater post-cracking flexural stiffness than normal-strength concrete (NSC) beams (El-Nemr et al. 2013). Because GFRP behaves in a linear elastic manner up to failure, the forces carried by stirrups crossing a diagonal shear crack are not evenly distributed. Instead, the force in each stirrup depends on where the crack intersects it and the amount of crack opening parallel to the stirrup The bent portions of GFRP stirrups are weaker than the straight segments and are more susceptible to localized failure at load levels lower than those expected based on the tensile capacity of the straight sections (Gomes et al. 2023; Resende et al. 2025).

The lower stiffness of GFRP reinforcement reduces the depth of the uncracked compression zone above the diagonal crack tip, which in turn limits the capacity of the concrete to transfer shear through this compressed region. Although earlier studies indicate that dowel action in GFRP bars may be reduced due to their anisotropic nature, the experimental findings of the present study revealed that increasing the bar diameter improved the dowel strength by about 32.7% (Nanni et al. 2014;

Table 3 Permissible crack width for GFRP-reinforced concrete members according to international codes (Deneko and Filaj 2022)

Code	Exposure classification	Maximum allowable crack width, W_{max} (mm)
JSCE	General service condition	0.51
CSA	Interior exposure	0.71
CSA	Exterior exposure	0.51
EUROCRETE	Waterproofing requirement	0.2
EUROCRETE	Aesthetic consideration	0.3–0.5
EUROCRETE	Structural performance requirement	0.5–1.0
ACI	Interior exposure	0.7
ACI	Exterior exposure	0.5
EUROCODE 2	General exposure condition	0.5

Resende et al. 2025). GFRP-reinforced beams exhibited reduced post-cracking stiffness because of both the lower modulus of elasticity and bond-slip effects. Consequently, larger deflections were observed in GFRP beams at comparable load levels. The addition of polypropylene fibres had only a marginal influence on post-cracking behaviour (Saikia et al. 2007).

5.5 Failure Modes and Ductility Considerations

Unlike steel reinforcement, FRP bars do not show yielding and fail in a sudden brittle manner. For this reason, most design codes recommend that FRP-reinforced beams be designed as over-reinforced sections, so that failure occurs by concrete crushing rather than by rupture of the FRP bars. Over-reinforced FRP beams usually exhibit smaller deflections and narrower cracks at service load levels compared to under-reinforced beams. This is because they have higher stiffness and the FRP bars experience lower stress under service conditions (Abdelkarim et al. 2019).

Ductility is referred as the ability of a structure to deform and absorb energy before failure. Since FRP bars does not exhibit yielding like steel, assessing ductility in FRP-reinforced concrete members is more challenging and requires special consideration. Therefore, assessing ductility is an important aspect of studying concrete beams reinforced with FRP bars. For steel-reinforced concrete, flexural ductility is commonly defined as the ratio of ultimate displacement to the displacement at steel yielding. In the case of FRP-reinforced members, different approaches have been proposed (Vijay et al. 2001; Naaman and Jeong 1995; Abdelkarim et al. 2019).

In general, studies have shown that the ductility of FRP-reinforced concrete beams is influenced more by the properties of the concrete than by the FRP reinforcement itself. Different methods have been proposed to evaluate ductility. One approach is the energy-based method, where ductility is defined as the ratio of the total energy absorbed up to failure to the energy absorbed during the elastic stage (Naaman and Jeong 1995). Another approach is the deformability-based method, which relates ductility to the ratio of curvature at ultimate load to the curvature at service load (Jaeger et al. 1997). The Canadian Highway Bridge Design Code (CSA S6-14) adopted this deformability concept and specifies minimum ductility values of 4.0 for rectangular sections and 6.0 for T-sections (Abdelkarim et al. 2019). In experimental observations, GFRP sheets showed relatively high ductility compared to other systems. After repair, several secondary cracks developed, with a total elongation of about 3 mm over an initial gauge length of 200 mm. The main failure mode observed was debonding of the GFRP sheets from the concrete surface (Sachdeva and Singh 2017).

6 Design Considerations and Code Provisions

The experimental results show that, unlike steel-reinforced members where design is usually governed by ultimate strength, GFRP-reinforced beams are often controlled by serviceability limits such as deflection and crack width.

6.1 Design Philosophy (Strength vs. Serviceability)

Even though the cracking load depends on the compressive strength of concrete, the ultimate strength of GFRP beams does not change with varying concrete strength. This is mainly because failure is governed by bond-slip rather than crushing of the concrete. At the ultimate load, surface deformations of GFRP bar was observed. Hence, the bond behaviour plays an important role in structural performance. At failure load, Polyester-based GFRP bars reach about 90% of their tensile capacity, whereas epoxy-based bars reached around 50% of their tensile strength. This suggests that the failure of GFRP-reinforced members is not only controlled by the tensile capacity of the bar but also the bond interaction and slip between the bar and concrete. This also influence both strength and serviceability.

Due to this behaviour, controlling crack width and deflection limits are important in the design of GFRP-reinforced beams. In many cases, serviceability requirements control the dimension of the member and reinforcement ratio rather than the ultimate strength. Therefore, proper attention should be given to bond characteristics, reinforcement detailing, and stiffness during design (Saikia et al. 2007). Figure 12 illustrates the elastic behavior of a GFRP-reinforced concrete section under flexure (a) A linear strain profile is shown across the depth of the section, indicating that plane sections remain plane. The top concrete is in compression while the GFRP reinforcement at the bottom experiences tensile strain (which is less than its ultimate strain (b) Stress distribution**:** The corresponding stress diagram in the elastic stage is presented. Concrete develops a triangular compressive stress block and the GFRP reinforcement carries tensile force. The internal forces are in equilibrium (Nkurunziza et al. 2005).

Design stress limits for GFRP bars at serviceability are significantly lower than ultimate tensile capacity to control crack width and creep rupture. Some guidelines limit service stress to about 20–30% of guaranteed tensile strength. Maximum allowable strain limits (e.g. 2000 micro strain in ISIS guidelines) are specified to control cracking. GFRP-reinforced beams generally have lower stiffness and develop wider cracks than steel-reinforced beams because of the lower modulus of elasticity of GFRP (Nkurunziza et al. 2005). However, increasing the axial reinforcement stiffness (EfAf) can effectively reduce deflection and crack width. A curvature limit of 0.005/d has been suggested as suitable for maintaining serviceability performance. According to CSA provisions, larger crack widths are permitted for FRP-reinforced members, up to 0.5 mm for exterior exposure and 0.7 mm for interior exposure. The

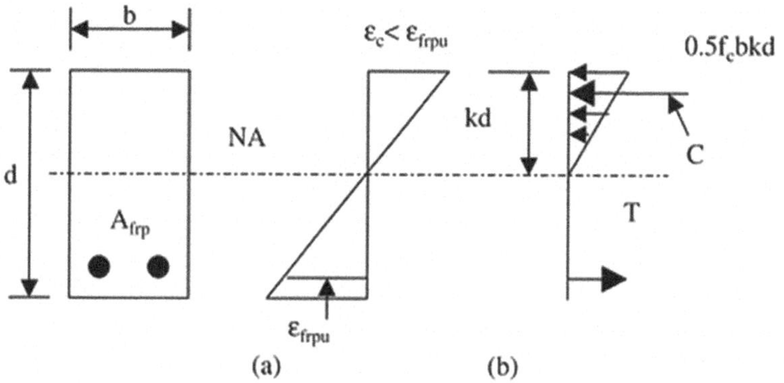

Fig. 12 Cross-Section of GFRP-Reinforced Concrete Member **a** strain distribution profiles **b** stress distribution under elastic conditions (Nkurunziza et al. 2005)

Table 4 Structural performance comparison: steel-RC versus GFRP-RC (El-Nemr et al. 2018)

Parameter	Steel-RC	GFRP-RC	Governing limit state
Failure Mode	Steel yielding	Concrete crushing	Compression-controlled design
Deflection	Lower	Higher	Serviceability governs
Crack width	Smaller	Larger allowable	0.5–0.7 mm permitted
Ductility	High	Limited (brittle)	No yielding plateau
Governing design check	Strength	Serviceability	Low modulus controls

ISIS guidelines recommend limiting the strain in FRP bars to about 2000 micro strain to control cracking (El-Nemr et al. 2018) (Table 4).

6.2 Safety Factors and Reduction Factors

In GB 50,608, an environmental reduction factor of 1.60 is recommended for GFRP bars used in marine conditions. Using this factor resulted in slightly conservative predictions, with the calculated residual flexural strength averaging about 86% of the experimental values. This indicates that residual strength data of GFRP bars can be effectively used to estimate the remaining flexural capacity of exposed beams, although refinement of the reduction factor is possible. The ACI 440.1R model also provided reasonable predictions of residual flexural strength, with an average predicted-to-test ratio of 0.91. Since this level of accuracy was acceptable, no modification to the ACI model was considered necessary (Wang et al. 2023).

Based on a comprehensive study of 576 beams and 288 columns hybrid steel-GFRP sections, two modifications (reduction/amplification) factors were developed: Fh to adjust section height and Fr to modify reinforcement area. These factors are used to compensate for the lower modulus of elasticity of GFRP and to ensure lateral performance comparable to steel-reinforced frames. Shear strength was verified using ACI 440-15 provisions, and flexural behavior governed the design (Youssef et al. 2017). Design codes such as ACI 440.1R-03, CAN/CSA-S6-00, and JSCE introduce environmental reduction factors and combined reduction factors to account for environmental degradation, sustained load, and creep effects. For GFRP bars, environmental reduction factors (CE) range from 0.7 to 0.8 depending on exposure conditions. When combined with sustained load effects, total reduction factors may range between 0.14 and 0.16. However, the paper reports that recent durability studies on newer generations of GFRP bars indicate that the strength reduction factors adopted in current codes are conservative and were originally based on early-generation products with limited long-term data (Nkurunziza et al. 2005). The balanced reinforcement ratio (ρfb) is defined based on FRP mechanical properties and concrete strength. ACI and CSA provisions recommend designing GFRP-reinforced sections as over-reinforced (ρf > ρfb) to ensure compression-controlled failure (concrete crushing). ACI specifies a minimum ρf/ρfb of 1.4. Ultimate concrete strain limits of 0.003 (ACI) and 0.0035 (CSA) are used in flexural capacity predictions. Experimental results showed good agreement between measured and predicted ultimate moments using ACI and CSA provisions (El-Nemr et al. 2018).

6.3 *International Guidelines (ACI, CSA, fib)*

The properties of FRP bars vary between manufacturers due to differences in fiber type, resin content, fiber–resin ratio, and production quality. Even minor manufacturing changes can affect strength, stiffness, and durability. Therefore, standardized testing methods are necessary to evaluate key properties such as tensile strength, modulus of elasticity, bond behavior, and long-term durability (Hajiloo et al. 2018). Unlike concrete and steel structures, which are supported by detailed fire design standards like EN 1992-1-2 for concrete and EN 1993-1-2 for steel, equivalent comprehensive fire design procedures for FRP structures are still limited (Correia et al. 2015). CSA S6-19 and CSA S807-19 strengthened this stress-controlled design philosophy by incorporating durability reduction factors and specific fatigue checks for GFRP-reinforced bridge structures. The publication of ACI 440.11–22 shows that GFRP is gradually being accepted as a regular reinforcement material in structural concrete. However, this document mainly focuses on limits related to sustained stress and serviceability, and it does not go deeply into fatigue design (Lee 2025). Several international codes, such as CSA S6-19, ACI 440.1R-15, CSA S806-12, and JSCE-97, also provide guidelines for calculating the required anchorage length of GFRP bars in concrete. These provisions help ensure adequate bond performance and effective stress transfer between the reinforcement and surrounding concrete (Wu et al. 2024). According to ACI 440.1R-06, the bond strength of FRP bars in concrete is estimated using a regression-based equation. This equation relates the normalized average bond stress to parameters such as concrete strength, concrete cover, and embedment length. The model expresses bond strength as a function of the square root of the concrete compressive strength, together with the ratios of concrete cover to bar diameter and embedment length to bar diameter.

The CSA S806-02 explains an expression to estimate the average bond strength of FRP bars embedded in concrete. This equation relates bond strength to key parameters such as concrete strength, bar diameter, and embedment conditions, and is intended to ensure adequate anchorage and safe stress transfer between the FRP reinforcement and surrounding concrete (Yan et al. 2016). ACI 440.1R-15 states that the anisotropic nature of FRP bars influences their structural behavior. Because their mechanical properties vary with direction, this affects their shear strength, dowel action, and bond performance in concrete (Arefian and Mostofinejad 2025). The load-carrying capacity and deflection of simply supported GFRP-reinforced concrete beams can be reasonably predicted using the equations provided in ACI 440.1R-06. These equations also give acceptable estimates for under-reinforced, continuously supported GFRP beams. However, the accuracy of these predictions becomes lower in the case of over-reinforced continuous GFRP beams (Habeeb and Ashour 2008). ACI 440.1R-06 estimates FRP bond strength based mainly on concrete strength, concrete cover, bar diameter, and embedment length. The equation was developed from 269 bond tests, mostly using GFRP bars. Different failure modes were observed, and bar surface type showed little influence, while the role of confinement requires further investigation. CSA S806-02 provides a more detailed bond strength model that considers concrete strength and cover, bar diameter, concrete density, bar location, fiber type, and surface condition. It includes modification factors to account for these parameters in practical design (Yan et al. 2016). All the major standards consider concrete strength, bar diameter, concrete cover, and bar location when calculating bond strength. However, embedment length is included only in ACI 440.1R-06. Bar surface profile did not show a clear influence on bond strength and requires further study. The Canadian and Japanese codes additionally consider factors such as fiber type and confinement from transverse reinforcement, which are not explicitly included in ACI 440.1R-06 (Yan et al. 2016) (Table 5).

6.4 Challenges in the Indian Context

Current Indian design standards are predominantly based on steel-reinforced concrete behaviour and it does not address mechanical characteristics of GFRP bars, particularly their low modulus of elasticity and bond-slip behaviour. In aggressive environments such as coastal and marine regions, where durability concerns favour GFRP, design based on crack width control is needed. Bond-slip behaviour, debonding of surface deformations, and differences in performance between epoxy- and polyester-based GFRP bars indicate that bond characteristics and binder type significantly influence structural response and failure mechanism of GFRP and should be included in the design codes (Saikia et al. 2007).

Table 5 Global design guidelines and standards for FRP reinforcement (Emparanza et al. 2022; Ortiz et al. 2023)

Organization	Key codes/guidelines	Main focus
AASHTO	GFRP-1 (LRFD Guide Specifications)	GFRP bridge deck and railing design
ACI	440.1R-15, 440.3R-12, 440.5–08, 440.6–08	Design, testing, and material specifications for FRP bars
CSA	S6-15, S806-12, S807-10; Design Manuals 3–5	Bridge and building design, material specs, rehabilitation, prestressing
ISO	14484, 25762, 10406–1	Performance guidelines, fire behavior, and test methods
ICC	AC454	Acceptance criteria for FRP reinforcement
fib	Bulletin 40	FRP reinforcement in RC structures
National codes	ISE (UK), SINTEF (Norway), CNR (Italy), STB (Belarus), ECP (Egypt), SAC GB/T series (China)	Structural design and material standards for FRP use

7 Sustainability and Life Cycle Perspective

7.1 Corrosion-Free Service Life

The use of GFRP as a substitute for conventional steel reinforcement has gained attention, primarily due to their ability to address corrosion-related durability concerns. Besides their resistance to corrosion, FRP rebar possess several advantages, such as high resistance to electrical and magnetic interference, high tensile strength, low self-weight, and good availability (Said et al. 2004; Razzaq et al. 2008).

A hybrid reinforcement approach that is combining GFRP and stainless steel (SS) rebar develop a corrosion-resistant reinforced concrete frame system. In this concept, SS rebar were placed in the plastic hinge regions to ensure adequate ductility and energy dissipation, while GFRP rebar were used in the remaining regions to enhance corrosion resistance. This hybrid strategy can achieve a corrosion-free structural system while maintaining acceptable structural performance (Youssef et al. 2017). Even in the bridges replacing steel reinforcement with GFRP bars can effectively eliminate corrosion problems (Mufti et al. 2010). Figure 13 shows the visible micro-cracks within the resin matrix between the circular glass fibers, indicating damage initiation. Such matrix cracking can reduce stiffness and facilitate moisture ingress, potentially leading to further degradation and fiber–matrix de-bonding.

GFRP bars does not show significant bond or strength reduction after freeze–thaw exposure. Strong alkaline environments reduced the strength of GFRP. Long-term field studies of 5–10 years under natural exposure conditions reported no obvious

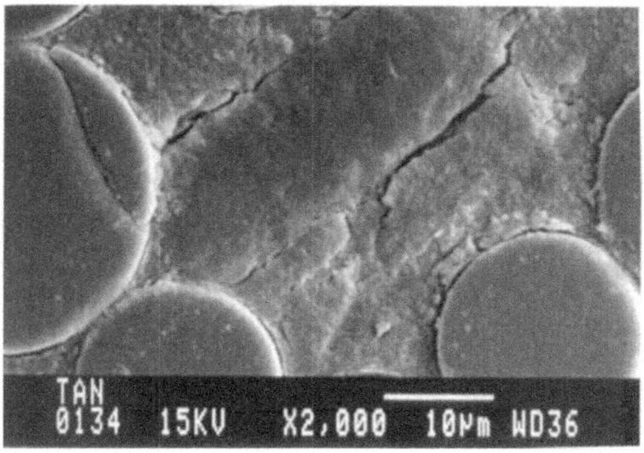

Fig. 13 SEM image shows matrix cracking in GFRP composite (Nkurunziza et al. 2005)

degradation in properly manufactured GFRP bars. However, very high temperatures (200–250 °C) significantly reduced bond strength (Yan et al. 2016). Recycled GFRP incorporation in concrete improves durability-related performance. Capillary water absorption was reported to reduce by up to ~ 80%, and total water absorption by up to ~ 90% at lower replacement levels. Drying shrinkage reductions of up to 50% were observed in some cases. Alkali–silica reaction (ASR) expansion remained below recommended standard limits, which was attributed to the protective resin coating surrounding the glass fibres (Tao et al. 2023).

7.2 Maintenance Reduction

Field investigations of real structures exposed to freeze–thaw, wet–dry cycles, and de-icing salts showed no clear degradation of GFRP reinforcement over several years. Proper manufacturing, especially adequate resin curing, plays an important role in improving durability. Alkali attack appears to cause more damage compared to freeze–thaw and wet–dry cycles (Yan et al. 2016). Life cycle cost analysis showed that GFRP-reinforced deck slabs have lower annual life cycle costs compared to steel-reinforced slabs. The annual worth of life cycle cost was about $176,000 for GFRP decks and $233,000 for steel decks, resulting in nearly 25% cost savings. The expected repair cycle for GFRP decks was also longer (50 years) compared to steel decks (25 years), further reducing maintenance frequency (Mufti et al. 2010). The non-corrodible nature of GFRP bars allowed elimination of the asphalt protection layer in a parking garage project, resulting in a 5% reduction in total project cost despite higher material cost of GFRP. This indicates potential reduction in protection and rehabilitation requirements associated with corrosion-prone steel reinforcement (El-Nemr et al. 2018).

7.3 Environmental Considerations

Concrete has a highly alkaline environment, which can influence FRP durability. Exposure to salt solutions, moisture changes, and temperature variations affects performance over time. Wet–dry cycles in salt conditions were reported to reduce strength gradually. Elevated temperatures are commonly used in laboratory studies to simulate long-term ageing effects (Yan et al. 2016). The alkaline environment (pH up to 13.5 in concrete pore solution) is identified as one of the most critical exposure conditions for glass fibers (Nkurunziza et al. 2005). Large amounts of glass waste are produced and disposed through landfilling or incineration, which raises environmental concerns. Reusing or recycled GFRP in concrete is considered as a sustainable choice for reducing this environmental imbalance. This helps to reduce the landfill requirement thereby lowering the environmental impact of concrete production, and decrease the need for natural aggregates.

Among all the available recycling methods, mechanical recycling requires less energy than thermal or chemical processes. Life cycle analysis studies shows that the total energy consumption and use of non-renewable resources is reduced by this recycling process. Hence, incorporating recycled GFRP in concrete offers a practical way to reduce environmental impact at the same time it supports circular economy principles (Tao et al. 2023).

8 Conclusions

- Compared with the normal steel, GFRP rebars have a high tensile strength and lower density. They also show a good resistance to chloride penetration, moisture, and chemical attack, which adds to the durability of reinforced concrete structures. But, their behaviour is different from steel reinforcement. They have a low modulus of elasticity and fail in a brittle manner without yielding or warning. Hence, deflection and crack width becomes the most important parameters for the design considerations.
- Another advantage of GFRP rebars is their low weight. They weigh about 75% less than a steel bar of the same diameter. This makes transportation easier and increases workability at the construction site. This light weight structure also reduces the effort required during installation and lowers risks of manual handling and hard labour. Since GFRP is non-corrosive, unlike steel, structures reinforced with GFRP achieves a longer service life, often estimated to have a life more than 100 years. The need for frequent repair or replacement is decreased and can be used as where more durable and economical concrete construction is required.
- However, the long -term performance of GFRP depends on several factors like moisture exposure, alkaline conditions, temperature, sustained loading, and the surface condition of the bars etc. The bonding between the GFRP and surrounding

concrete is also important, and proper detailing is necessary to prevent undesirable failures. Even though international design codes are available, practically challenges exist especially when used for under local construction.

References

Abdelkarim OI, Ahmed EA, Mohamed HM, Benmokrane B (2019) Flexural strength and serviceability evaluation of concrete beams reinforced with deformed GFRP bars. Eng Struct 186:282–296. https://doi.org/10.1016/j.engstruct.2019.02.024

Aly MYE, El-Hacha R (2009) Strength evaluation of steel-concrete composite girders strengthened with prestressed FRP laminates

Achillides Z, Pilakoutas K (2004) Bond behaviour of fiber reinforced polymer bars under direct pullout conditions. J Compos Constr 8:173–181. https://doi.org/10.1061/(ASCE)1090-0268(2004)8:2(173)

ACI Committee 440 (2015) Guide for the design and construction of structural concrete reinforced with FRP Bars (ACI 440.1R-15). American Concrete Institute, Farmington Hills, MI,USA. https://www.proquest.com/docview/2407605255?pq-origsite=gscholar&fromopenview=true&sourcetype=Dissertations%20&%20Theses

Adam MA, Said M, Mahmoud AA, Shanour AS (2015) Analytical and experimental flexural behavior of concrete beams reinforced with glass fiber reinforced polymers bars. Constr Build Mater 84:354–366. https://doi.org/10.1016/j.conbuildmat.2015.03.057

Alaoud L, Al-Salloum Y, Abbas H (2021) Experimental investigation for GFRP rebar couplers for reinforced concrete. J King Saud Univ-Eng Sci 33(2):104–110. https://doi.org/10.1016/j.jksues.2020.04.002

Alhayek A, Syamsir A, Supian ABM, Usman F, Asyraf MRM, Atiqah MA (2022) Flexural creep behaviour of pultruded GFRP composites cross-arm: a comparative study on the effects of stacking sequence. Polymers 14(7):1330. https://doi.org/10.3390/polym14071330

Ali HH, Said AMI (2022) Flexural behavior of concrete beams with horizontal and vertical openings reinforced by glass-fiber-reinforced polymer (GFRP) bars. J Mech Behav Mater 31(1):407–415. https://doi.org/10.1515/jmbm-2022-0045/html

Alsayed SH (1998) Flexural behaviour of concrete beams reinforced with GFRP bars. Cem Concr Compos 20(1):1–11. https://pdf.sciencedirectassets.com/271643/1-s2.0-S0958946500X00169/1-s2

Alsuhaibani E, Alturki M, Alogla SM, Alawad O, Alkharisi MK, Bayoumi E, Aldukail A (2024) Compressive and bonding performance of GFRP-reinforced concrete columns. Buildings 14(4):1071. https://doi.org/10.3390/buildings14041071

Alves J, El-Ragaby A, El-Salakawy E (2010) Durability of GFRP bars' bond to concrete under different loading and environmental conditions. J Compos Constr 15(3):249e62. https://doi.org/10.1061/(ASCE)CC.1943-5614.0000161

Al-Zahrani MM (2024) Durability of concrete-embedded GFRP bars after 20 years of tidal zone exposure: correlation with accelerated aging tests. Case Stud Constr Mater 21:e03435. https://doi.org/10.1016/j.cscm.2024.e03435

Ananthkumar M, Mini KM, Prakash C, Sharma SV, Krishnaa AB (2020) Study on the efficiency of CFRP and GFRP in corrosion resistance of rebar embedded in concrete. In: IOP conference series: materials science and engineering, vol. 872, no 1. IOP Publishing, p 012137. https://doi.org/10.1088/1757-899X/872/1/012137/meta

Arefian B, Mostofinejad D (2025) Bond strength and load-carrying capacity of GFRP rebars embedded in concrete: an experimental and analytical study. Constr Build Mater 479:141512. https://doi.org/10.1016/j.conbuildmat.2025.141512

Ashby MF, Jones DRH (1997) Engineering materials 2: an introduction to microstructures, processing, and design. Pergamon Press, Oxford. https://cir.nii.ac.jp/crid/1970867909884625550

Ashrafi H, Bazli M, Oskouei AV (2017) Enhancement of bond characteristics of ribbed-surface GFRP bars with concrete by using carbon fiber mat anchorage. Constr Build Mater 134:507–519. https://doi.org/10.1016/j.conbuildmat.2016.12.083

Aybar K, Eryilmaz Yildirim M, Canbaz M (2025) Effect of high temperature on the mechanical behavior of glass fiber-reinforced polymer composite bars under compressive loading. J Mater Civ Eng 37(9):04025302. https://doi.org/10.1061/JMCEE7.MTENG-20025

Azizinamini A, Stark M, Roller JJ, Ghosh S (1993) Bond performance of reinforcing bars embedded in high-strength concrete. ACI Struct J 90(5). https://www.concrete.org/publications/internationalconcreteabstractsportal.aspx?m=details&id=3951

Baena M, Torres LL, Turon A, Barris C (2009) Experimental study of bond behavior between concrete and FRP bars using a pull-out test. Compos: Part B 40:784–797. https://doi.org/10.1016/j.compositesb.2009.07.003

Bank LC (2006) Composites for construction: structural design with FRP materials. John Wiley & Sons, New Jersey, USA. https://books.google.co.in/books

Benmokrane B, Chaallal O, Masmoudi R (1995) Glass fibre reinforced plastic (GFRP) rebars for concrete structures. Constr Build Mater 9(6):353–364. https://doi.org/10.1016/0950-0618(95)00048-8

Benmokrane B, Chaallal O, Masmoudi R (2002) Glass fibre reinforced plastic (GFRP) rebars for concrete structures. Constr Build Mater 16(3):143–153. https://doi.org/10.1016/0950-0618(95)00048-8

Benmokrane B, Ali AH, Mohamed HM, ElSafty A (2017a) Laboratory assessment and durability performance of vinyl-ester, polyester, and epoxy glass-FRP bars for concrete structures. Compos Part B 114:163–174. https://doi.org/10.1016/j.compositesb.2017.02.002

Benmokrane B, El-Safty A, Hassan M (2017b) Performance of GFRP bar-reinforced concrete bridge decks in North America. PCI J 62(3):42–56

Benmokrane B, Tighiouart B. Bond strength and load distribution of composite GFRP reinforcing bars in concrete. ACI Mater J 93(3). https://www.concrete.org/publications/internationalconcreteabstractsportal.aspx?m=details&ID=9810&m=details&ID=9810

Brown VL, Bartholomew CL (1993) FRP reinforcing bars in reinforced concrete members. ACI Mater J 90(1). https://www.researchgate.net/profile/Vicki-Brown-2/publication/279574868_

Brown JR, Mathys Z (1997) Reinforcement and matrix effects on the combustion properties of glass reinforced polymer composites. Compos A Appl Sci Manuf 28(7):675–681. https://doi.org/10.1016/S1359-835X(97)00018-3

Cai YJ, Xiao SH, Chen YW, Huang ZR, Lin JX, Guo YC, Xie ZH et al (2023) Tensile behavior and durability prediction of GFRP-steel composite bars under chloride environments. J Mater Res Technol 23:5746–5759.https://doi.org/10.1016/j.jmrt.2023.02.165

Chaallal O, Benmokrane B (1993) Pullout and bond of glass-fibre rods embedded in concrete and cement grout. Mater Struct 26(3):167e75. https://doi.org/10.1007/BF02472934

Chen Z, Li S, Zhou J, Xu R, Dai S (2022) Flexural behavior of GFRP bars reinforced seawater sea sand concrete beams exposed to marine environment: experimental and numerical study. Constr Build. Mater. 349:128784. https://doi.org/10.1016/j.conbuildmat.2022.128784

Correia JR, Bai Y, Keller T (2015) A review of the fire behaviour of pultruded GFRP structural profiles for civil engineering applications. Compos Struct 127:267–287. https://doi.org/10.1016/j.compstruct.2015.03.006

Cosenza E, Manfredi G, Realfonzo R (1997) Behavior and modeling of bond of FRP rebars to concrete. J Compos Constr 1(2):40e51. https://doi.org/10.1061/(ASCE)1090-0268(1997)1:2(40)

Czarnecki L, Garbacz A, Wang R, Frigione M, Aguiar JB (Eds) (2024) Concrete-polymer composites in circular economy: proceedings of the 17th international congress on polymers in concrete (ICPIC 2023). Springer Nature. https://doi.org/10.1007/978-3-031-72955-3_41

D'antino T, Pisani MA, Poggi C (2018) Effect of the environment on the performance of GFRP reinforcing bars. Compos B Eng 141:123–136. https://doi.org/10.1016/j.compositesb.2017.12.037

Deneko E, Filaj E (2022) A study on the cracking behavior of gfrp reinforced concrete beams. https://pdfs.semanticscholar.org/ca20/83d1a556e5cd841967636355912bb879eb00.pdf

Devaraj R, Olofinjana A, Gerber C (2023) On the factors that determine the bond behaviour of GFRP bars to concrete: an experimental investigation. Buildings 13(11):2896. https://doi.org/10.3390/buildings13112896

Dwight DW (2000) Glass fiber reinforcements. In: Kelly A, Zweben C (eds) Comprehensive composite materials. Elsevier. https://irepository.uniten.edu.my/entities/publication/e1053cf1-daee-492e-9116-937b5ff62087

Ehsani M, Saadatmanesh H, Tao S (1993) Bond of GFRP rebars to ordinary-strength concrete. ACI Spec Publ 138. https://www.concrete.org/publications/internationalconcreteabstractsportal.aspx?m=details&ID=3930

El-Hassan H, El Maaddawy T (2019) Microstructure characteristics of GFRP reinforcing bars in harsh environment. Adv Mater Sci Eng 2019(1):8053843. https://doi.org/10.1155/2019/8053843

Elkafrawy M, Gowrishankar P, Aswad NG, Alashkar A, Khalil A, AlHamaydeh M, Hawileh R (2024) GFRP-reinforced concrete columns: state-of-the-art, behavior, and research needs. Buildings 14(10):3131. https://doi.org/10.3390/buildings14103131

El-Nemr A, Ahmed EA, El-Safty A, Benmokrane B (2018) Evaluation of the flexural strength and serviceability of concrete beams reinforced with different types of GFRP bars. Eng Struct 173:606–619. https://doi.org/10.1016/j.engstruct.2018.06.089

El-Nemr EA, Ahmed B (2013) Benmokrane flexural behavior and serviceability of normal-and high-strength concrete beams reinforced with glass fiber-reinforced polymer bars ACI Struct J 110(6):1077. https://www.researchgate.net/profile/Amr-Elnemr/publication/256287944_Flexural_Behavior_and_Serviceability_of_Normal-_and_High-Strength_Concrete_Beams_Reinforced_with_Glass_Fiber-Reinforced_Polymer_Bars

Emparanza AR, Kampmann R, De Caso F, Morales C, Nanni A (2022) Durability assessment of GFRP rebars in marine environments. Constr Build Mater 329:127028. https://doi.org/10.1016/j.conbuildmat.2022.127028

Fergani H, Di Benedetti M, Oller CM, Lynsdale C, Guadagnini M (2018) Durability and degradation mechanisms of GFRP reinforcement subjected to severe environments and sustained stress. Constr Build Mater 170:637–648. https://doi.org/10.1016/j.conbuildmat.2018.03.092

Gao H, Sun Y, Jian J, Dong Y, Liu H (2023) Study on mechanical properties and application in communication pole line engineering of glass fiber reinforced polyurethane composites (GFRP). Case Stud Const Mater 18:e01942. https://doi.org/10.1016/j.cscm.2023.e01942

Gaurav A, Singh KK (2018) Fatigue behavior of FRP composites and CNT-embedded FRP composites: a review. Polym Compos 39(6):1785–1808. https://doi.org/10.1002/pc.24177

Ghasemzadeh F, Manafpour A, Sajedi S, Shekarchi M, Hatami M (2016) Predicting long-term compressive creep of concrete using inverse analysis method. Constr Build Mater 124:496–507. https://doi.org/10.1016/j.conbuildmat.2016.06.137

Gomes TA, Resende TL, Cardoso DCT (2023) Shear-transfer mechanisms in reinforced concrete beams with GFRP bars and basalt fibers (DOI:) Eng Struct, 289(2023), Article 116299. https://doi.org/10.1016/j.engstruct.2023.116299

Habeeb MN, Ashour AF (2008) Flexural behavior of continuous GFRP reinforced concrete beams. J Compos Construct 12(2):115–124. https://ascelibrary.org/doi/epdf/10.1061

Hajiloo H, Green MF, Gales J (2018) Mechanical properties of GFRP reinforcing bars at high temperatures. Constr Build Mater 162:142–154. https://doi.org/10.1016/j.conbuildmat.2017.12.025

Hollaway LC (2010) A review of the present and future utilisation of FRP composites in the civil infrastructure with reference to their important in-service properties. Constr Build Mater 24(12):2419–2445. https://doi.org/10.1016/j.conbuildmat.2010.04.062

Hu JY, Zhang SS, Chen E, Li WG (2022) A review on corrosion detection and protection of existing reinforced concrete (RC) structures. Constr Build Mater 325:126718. https://doi.org/10.1016/j.conbuildmat.2022.126718

Hu X, Xiao J, Zhang K, Zhang Q (2023) Evaluation of cracking and deflection of GFRP bar reinforced recycled concrete beams with seawater and sea sand. J Build Eng 80:108092. https://doi.org/10.1016/j.jobe.2023.108092

Jabbar SA, Farid SB (2018) Replacement of steel rebars by GFRP rebars in the concrete structures. Karbala Int J Mod Sci 4(2):216–227. https://doi.org/10.1016/j.kijoms.2018.02.002

Jaeger GL, Tadros G, Mufti AA (1997) The concept of the overall performance factor in rectangular-section reinforced concrete beams. In: Proceeding of 3rd int symp on non-metallic (FRP) reinforcement for concrete structures, vol 2, Sapporo, Japan. pp 551–558

Kamal ASM, Boulfiza M (2011) Durability of GFRP rebars in simulated concrete solutions under accelerated aging conditions. J Compos Constr 15(4):473–481. https://doi.org/10.1061/(ASCE)CC.1943-5614.0000168

Karataş MA, Gökkaya H (2018) A review on machinability of carbon fiber reinforced polymer (CFRP) and glass fiber reinforced polymer (GFRP) composite materials. Defence Technol 14(4):318–326. https://doi.org/10.1016/j.dt.2018.02.001

Lee SY (2025) Fatigue behaviors of high-speed track slabs reinforced by GFRP composite rebar: full-scale experimental verification. J Compos Sci 9(11):597. https://doi.org/10.3390/jcs9110597

Li VC, Wang S (2002) Flexural behaviors of glass fiber-reinforced polymer (GFRP) reinforced engineered cementitious composite beams. Mater J 99(1):11–21. https://www.researchgate.net/profile/Victor-Li-15/publication/280229572_

Li X, Zhang W, Zhang C, Liu J, Li L, Wang S (2024) Flexural behavior of GFRP and steel bars reinforced lightweight ultra-high performance fiber-reinforced concrete beams with various reinforcement ratios. In: Structures, vol 70. Elsevier, pp 107897 https://doi.org/10.1016/j.istruc.2024.107897

Lim S, Akiyama M, Frangopol DM (2016) Assessment of the structural performance of corrosion-affected RC members based on experimental study and probabilistic modeling. Eng Struct 127:189–205. https://doi.org/10.1016/j.engstruct.2016.08.040

Liu Y, Zhang HT, Zhao HH, Lu L, Han MY, Wang JC, Guan S (2021) Experimental study on mechanical properties of novel FRP bars with hoop winding layer. Adv Mater Sci Eng (1):9554687. https://www.researchgate.net/publication/353696190_

Makitani E, Irisawa I, Nishiura N (1993) Investigation of bond in concrete member with fiber reinforced plastic bars. ACI Spec Publ 138. https://www.concrete.org/publications/internationalconcreteabstractsportal.aspx?m=details&id=3929

Maranan GB, Manalo AC, Benmokrane B, Karunasena W, Mendis P, Nguyen TQ (2019) Flexural behavior of geopolymer-concrete beams longitudinally reinforced with GFRP and steel hybrid reinforcements Eng Struct 182:141–152, https://doi.org/10.1016/j.engstruct.2018.12.073

Mouritz AP, Gibson AG (2006) Fire properties of polymer composite materials. Dordrecht: Springer. https://doi.org/10.1007/978-1-4020-5356-6_8

Movahedi-Rad AV, Keller T, Vassilopoulos AP (2019) Creep effects on tension-tension fatigue behavior of angle-ply GFRP composite laminates. Int J Fatigue 123:144–156. https://doi.org/10.1016/j.ijfatigue.2019.02.010

Mufti A, Bakht B, Tadros G (2010) GFRP reinforced corrosion-free bridge deck slabs based on ISIS Winnipeg principles. Adv Struct Eng 13(5):773–782. https://doi.org/10.1260/1369-4332.13.5.773

Naaman AE, Jeong SM (1995) Structural ductility of concrete beams prestressed with FRP tendons In: Taerwe L (ed) Proceeding of 2nd int RILEM symp (FRPRXS-2), non-metric (FRP) reinforcement for concrete structures, RILEM, E & FN Spon, London, pp 379–386

Nanni A, De Luca A, Jawaheri Zadeh H (2014) Reinforced concrete with FRP bars: mechanics and design. CRC Press. https://www.concrete.org/publications/internationalconcreteabstractsportal.aspx?m=details&id=3951

Natarajan V, Samraj R, Duraivelu J, Paulraj P (2021) Experimental investigation on mechanical properties of A/GFRP, B/GFRP and AB/GFRP polymer composites. Mater Plast 58(4):28–36. https://doi.org/10.37358/Mat.Plast.1964

Nguyen MH, Pham VN, Phan HN, Do VH, Huynh PN, Mai TTT (2025) Gene-expression programming-based models for comprehensive evaluation of bond strength and critical slip of various types of FRP bars in concrete. Struct Concrete 26(3):3700–3719. https://www.researchgate.net/publication/388062351_

Nkurunziza G, Debaiky A, Cousin P, Benmokrane B (2005) Durability of GFRP bars: a critical review of the literature. Prog Struct Eng Mater 7(4):194–209. https://doi.org/10.1002/pse.205

Ohlemiller TJ, Shields JR (1999) The effect of surface coating on fire growth over composite materials in a corner configuration. Fire Saf J 32(2):173–193. https://doi.org/10.1016/S0379-7112(98)00031-9

Olczyk N, Błyszko J, Techman M, Zielinski A (2019) Stress analysis of slabs reinforced with GFRP rebar. In: IOP conference series: materials science and engineering, vol 471, no 3. IOP Publishing, p 032056. https://doi.org/10.1088/1757-899X/471/3/032056/meta

Ortiz JD, Khedmatgozar Dolati SS, Malla P, Nanni A, Mehrabi A (2023) FRP-reinforced/strengthened concrete: state-of-the-art review on durability and mechanical effects. Materials 16(5):1990. https://doi.org/10.3390/ma16051990

Pecce M, Manfredi G, Realfonzo R, Cosenza E (2001) Experimental and analytical evaluation of bond properties of GFRP bars. J Mater Civ Eng 13(4):282e90. https://doi.org/10.1061/(ASCE)0899-1561(2001)13:4(282)

Qureshi J (2022) A review of fibre reinforced polymer structures. Fibers 10(3):27. https://doi.org/10.3390/fib10030027

Rajak DK, Wagh PH, Linul E (2021) Manufacturing technologies of carbon/glass fiber-reinforced polymer composites and their properties: a review. Polymers 13(21):3721. https://doi.org/10.3390/polym13213721

Razzaq Z, Mojtaba S, Prabhakaran R (2008) LRFD approach for FRP channel section columns. J Reinf Plast Compos 28(12):1409–1415. https://doi.org/10.1177/0731684408089488

Resende TL, Pacheco DD, Cardoso DC, Noël M (2025) Analysis of the shear behavior of GFRP reinforced concrete beams through crack kinematics and shear transfer mechanisms. Eng Struct 340:120778. https://doi.org/10.1016/j.engstruct.2025.120778

Robert M, Benmokrane B (2010) Physical, mechanical, and durability characterization of preloaded GFRP reinforcing bars. J Compos Constr 14(4):368–375. https://doi.org/10.1061/(ASCE)CC.1943-5614.0000104

Rodrigues R, Gaboreau S, Gance J, Ignatiadis I, Betelu S (2021) Reinforced concrete structures: a review of corrosion mechanisms and advances in electrical methods for corrosion monitoring. Constr Build Mater 269:121240. https://doi.org/10.1016/j.conbuildmat.2020.121240

Saad S, Polak MA (2025) Bond behavior of glass fiber-reinforced polymer (GFRP) bars embedded in concrete: a review. Materials 18(14):3367. https://pmc.ncbi.nlm.nih.gov/articles/PMC12300568/

Sachdeva R, Singh H (2017) A review of application of GFRP in shear and flexure to strengthen the reinforced concrete beam. Int J Eng Res & Technol (IJERT) Ncietm 5(11). https://www.ijert.org/a-review-of-application-of-gfrp-in-shear-and-flexure-to-strengthen-the-reinforced-concrete-beam

Said AM, Nehdi ML. Use of FRP for RC frames in seismic zones: part I. Evaluation of FRP beam-column joint rehabilitation techniques. Appl Compos Mater 11(4):205–226. https://doi.org/10.1023/B:ACMA.0000035462.41572.7a

Saikia B, Kumar P, Thomas J, Rao KN, Ramaswamy A (2007) Strength and serviceability performance of beams reinforced with GFRP bars in flexure. Constr Build Mater 21(8):1709–1719. https://doi.org/10.1016/j.conbuildmat.2006.05.021

Sanjay MR, Arpitha GR, Yogesha B (2015) Study on mechanical properties of natural-glass fibre reinforced polymer hybrid composites: a review. Mater Today: Proc 2(4–5):2959–2967. https://doi.org/10.1016/j.matpr.2015.07.264

Sasikumar P, Manju R (2024) Flexural behaviour of reinforced concrete beams reinforced with glass fibre reinforced polymer (GFRP) bars: experimental and analytical study. Asian J Civ Eng 25:3623–3636. https://doi.org/10.1007/s42107-024-01000-4

Sateesh N, Rao PS, Ravishanker DV, Satyanarayana K (2015) Effect of moisture on GFRP composite materials. Mater Today: Proc 2(4–5):2902–2908. https://doi.org/10.1016/j.matpr.2015.07.252

Scudamore MJ (1994) Fire performance studies on glass-reinforced plastic laminates. Fire Mater 18(5):313–325. https://doi.org/10.1002/fam.810180507

Shabani H, Asadian A, Galal K (2025) Flexural and serviceability behaviour of macro-synthetic fibre-reinforced concrete beams reinforced with GFRP bars. Constr Build Mater 494:143217

Tao Y, Hadigheh SA, Wei Y (2023, July) Recycling of glass fibre reinforced polymer (GFRP) composite wastes in concrete: a critical review and cost benefit analysis. In: Structures, vol 53, pp 1540–1556. Elsevier. https://doi.org/10.1016/j.istruc.2023.05.018

Tepfers R (2006) Bond clause proposals for FRP bars/rods in concrete based on CEB/FIP Model Code 90. Part 1: design bond stress for FRP reinforcing bars. Struct Concrete 7(2):47–55. https://doi.org/10.1680/stco.2006.7.2.47

Tran H, Nguyen-Thoi T, Dinh HB (2025) State-of-the-art review of studies on the flexural behavior and design of frp-reinforced concrete beams. Materials 18(14):3295. https://doi.org/10.3390/ma18143295

Vieille B, Albouy W, Taleb L (2014) About the creep-fatigue interaction on the fatigue behaviour of off-axis woven-ply thermoplastic laminates at temperatures higher than Tg. Compos B Eng 58:478–486. https://doi.org/10.1016/j.compositesb.2013.11.005

Vijay PV, Ganga Rao HV (2001) Bending behavior and deformability of glass fiber-reinforced polymer reinforced concrete members Struct J, 98(6):834–842. https://www.concrete.org/publications/internationalconcreteabstractsportal.aspx?m=details&ID=10750

Wang Z, Xie J, Mai Z, Liu P, Lu Z, Li L (2023) Durability of GFRP bar-reinforced seawater–sea sand concrete beams: coupled effects of sustained loading and exposure to a chloride environment. Eng Struct 283:115814. https://doi.org/10.1016/j.engstruct.2023.115814

Wen S, Cao M (2023) Review on degradation behavior of fiber-reinforced polymer bars in marine environments. J Mater Sci 58(24):9861–9893. https://doi.org/10.1007/s10853-023-08667-1

Won JP, Lee SJ, Kim YJ, Jang CI, Lee SW (2008) The effect of exposure to alkaline solution and water on the strength–porosity relationship of GFRP rebar. Compos B Eng 39(5):764–772. https://doi.org/10.1016/j.compositesb.2007.11.002

Wong EH, Mai YW (2014) A unified equation for creep-fatigue. Int J Fatigue 68:186–194. https://doi.org/10.1016/j.ijfatigue.2014.05.004

Wu Z, Wang X, Zhao X, Noori M (2014) State-of-the-art review of FRP composites for major construction with high performance and longevity. Int J Sustain Mater Struct Sys 1(3):201–231. https://www.inderscienceonline.com/doi/abs/10.1504/IJSMSS.2014.062757

Wu W, He X, Yang W, Tang Z, Wang H, Zhou M, Wei B, He J (2024) Investigation of long-term degradation of the interface in the anchorage zone of GFRP-RC beams. Case Stud Constr Mater 21:e03399. https://doi.org/10.1016/j.cscm.2024.e03399

Wu G, Wang X, Wu Z, Dong Z (2020) Durability of basalt fiber reinforced polymer (BFRP) bars in concrete. Compos Struct 236:111849

Wu W, He X, Yi Z, Zhu Z, He J, Wang W, Zhao C (2022) Flexural fatigue behaviors of high-content hybrid fiber-polymer concrete. Constr Build Mater 349:128772. https://doi.org/10.1016/j.conbuildmat.2022.128772

Yan F, Lin Z, Yang M (2016) Bond mechanism and bond strength of GFRP bars to concrete: a review. Compos B Eng 98:56–69. https://doi.org/10.1016/j.compositesb.2016.04.068

Youssef MA, Meshaly ME, Elansary AA (2017) Ductile corrosion-free GFRP-stainless steel reinforced concrete elements. Compos Struct 182:124–131. https://doi.org/10.1016/j.compstruct.2017.09.037

Zhang H, Davalos JF, Ray I (2021) Long-term durability and degradation modeling of FRP reinforcement in concrete. Compos Struct 256:113040

Zuo J, Darwin D (2000) Splice strength of conventional and high relative rib area bars in normal and high-strength concrete. ACI Struct J 97(4):630–641. https://www.researchgate.net/profile/David-Darwin/publication/279587636

Application of Artificial Intelligence for Rehabilitation of Infrastructures Affected by Vibrations

Prafulla Parlewar

Abstract The application of artificial intelligence in the rehabilitation of infrastructures affected by vibration examines the influence of vibration on structures and the essential strategies for rehabilitation. The study showcases a project where suboptimal design affected an air fan cooler support structure, leading to high-frequency vibrations. The study primarily focuses on developing an artificial neural network (ANN) to predict fault frequency and estimate structural failures. The study explores the application of artificial intelligence in the rehabilitation of structures impacted by vibrations. How are artificial neural networks developed for vibration predictions? Additionally, the investigation explores the various methods of rehabilitation for these structures. The rehabilitation process examines the simulation of the project for both pre-rehabilitation and post-rehabilitation conditions. The outcome provided a potential solution for designing new braces aimed at controlling the overall vibration in air fan coolers.

Keywords Artificial neural network · Steel structures · Rehabilitation · Vibration · Infrastructure

1 Introduction

The application of artificial intelligence (AI) for vibration-damaged structure investigates the use of artificial neural network (ANN) models for prediction and data training. The project involves a framework for air fan coolers that experience significant vibrations, leading to structural distortion. The rehabilitation process specifically examines the application of the machine learning model in the structural rehabilitation context. What are the appropriate methods for analyzing the vibration model? What would be an appropriate solution to reduce the vibration in the structure? The project structure experienced vibrations from the air fan, leading to structural degradation and performance failure. So far, AI has shown that it can handle large

P. Parlewar (✉)
School of Planning and Architecture, New Delhi, India
e-mail: prafulla.parlewar@spa.ac.in

João M. P. Q. Delgado (ed.), *Advanced Materials, Corrosion Science, and Rehabilitation of Concrete Infrastructure*, Building Pathology and Rehabilitation 29,
https://doi.org/10.1007/978-3-032-28851-6_6

datasets well and use pattern recognition and predictive analysis to find damage in systems that are affected by high-frequency vibrations. This is changing how similar structures are fixed. AI opens new possibilities for structural health monitoring and rehabilitation. AI technologies in the form of artificial neural networks have shown a lot of promise in predicting damage to structures caused by vibrations and the steps that need to be taken to fix them. A steel framework supports an extensive arrangement of air fans in the project. The diameter of each air fan was 13 ft, and the structure was designed with a width of 9.0 m, featuring a continuous length to accommodate twenty-eight fans arranged in longitudinal directions. The entire structure was upheld by a concrete pedestal measuring 25 mm in thickness. The elevation of the structure measured 4.5 m from the apex of the fan. The structure features movable motors, a trolley, a ladder, vibration control, and catwalks designed for maintenance purposes. ISMB500 at the top and ISMB200 at the bottom provide horizontal support for the structure. The vertical members consist of ISMB400. The previous data regarding vibration was located for the structure. The research analyzed this historical data to determine how the motor drive end vibrated in axial directions. Further, this research collected vibration data at the non-directional end of the motor with high resolution, concentrating on horizontal directions. The research also collected vibration data in vertical orientations. The observations were conducted using motion amplification videos. The preliminary analysis revealed significant vibration and bending within the structure. The second stage of data collection involved the diagnosis of the complete structure. The project consists of an extensive arrangement of air fans, all supported by a steel framework, featuring a total of 28 fans. The approach implemented for the restoration of the structure was outlined as follows: (1) Geometric studies, (2) Structural testing, (3) Software modeling, and (4) Prediction modeling in ANN. The geometric investigations encompassed the documentation and measurement of structural damages via visual analyses. This encompassed the identification of damages, corrosion, defects in welded joints, joinery, and so forth. The structural testing involved the use of rebound hammer tests to assess the overall strength of the concrete. In order to comprehend the deflections of a structure influenced by vibration, it is crucial to grasp the structural integrity of both the foundation and the superstructure. The subsequent phase of the procedure involved the collection of vibration data, which is crucial for modeling in STAAD and KNIME. We identified the rehabilitation methods for the structure in the final stage.

The literature review on the research has revealed that there are many research work in field of vibration analysis. There has been a Bayesian deep learning approach for random vibration analysis of bridges subjected to vehicle dynamic integration by Li et al. (2022). In research of prediction of bolt loosing using vibration analysis and machine learning, one study has utilized the machine learning algorithm for predicting bolt loosening by analyzing the vibration data (Pham and Yang 2022). Along with ANN convolutional neural networks in analyzing vibration signals for tasks like surface roughness estimation and fault diagnosis is a deep learning approach for vibration signal analysis (Chen and Lee 2021). One other study has built upon previous research that utilizes a vibration based machine learning approach for diagnosis of rotary-related faults in rotating machinery (Almutairi and Sinha 2023). In research by Costa et al. (2024), an approach for supervised learning is illustrated by the author. The study monitors low-speed bearings using vibration analysis is more challenging due to the low energy of the interest information, as the vibration signal is more susceptible to interference from other sources (Costa et al. 2024). In some approached the dynamic analysis of the drill string vibration enhanced by machine learning models is illustrated. This research utilizes machine learning methods to predict the torsional and axial vibration of the drill string (Chen et al. 2024). The analysis of vibration data using machine learning promises a significant reduction in analysis effort and improved diagnostic accuracy for unbalance detection in rotating machinery (Mey et al. 2020). An Ensemble of Convolution-Based Methods for Fault Detection Using Vibration Signals by Lee et al. (2023) proposes an ensemble of convolution-based methods for fault detection using multivariate time series of vibration signals collected from planetary gearboxes. Another study on Learning to Predict Structural Vibrations presents a benchmark on predicting the vibration of harmonically excited plates using deep learning models (Van Delden et al. 2023). A study introduces Resoformer, a transformer-based model for predicting torsional resonance in electric vehicles (Kuniyoshi and Sawada 2024). Sharma et al. study highlights the application of ANN techniques in condition monitoring of industrial fans using vibration signals, emphasizing improved fault quantification and classification (Sharma and Sharma 2024). The research by Vallim et al. presents a methodology using ANN to predict motor failure time based on vibration data, allowing for simulation of vibration behaviors and effective maintenance planning (Vallim and de Souza 2019). The study by Hakim et al. applies ANN to vibration test data for detecting damage in steel girder bridges, demonstrating the model's effectiveness in identifying damage location and extent (Hakim and Abdul Razak 2011).

Papanikolaou et al. has indicated that a comprehensive assessment is necessary for developing tailored rehabilitation approaches to address the effects of vibrations on structural integrity (Papanikolaou and Kartsaklis 2020). Mohammed et al. discusses innovative rehabilitation methods specifically designed for bridges experiencing excessive dynamic loads from traffic (Mohammed 2021). Further in research by Brzyski, the study evaluates the potential damage caused by vibrations and suggests targeted rehabilitation strategies to enhance structural resilience (Brzyski 2019). Khan et al. demonstrates that the presence of cracks in steel structures leads to changes in mechanical properties, resulting in decreased modal frequencies. The research explains linear trend observed between frequency reduction and increasing crack depth, aiding in structural health monitoring (Khan et al. 2020). Lokade discusses the importance of designing steel frame structures to accommodate dynamic loads from rotating machinery, emphasizing the need for vibration control measures to prevent structural issues (Lakade 2019). The research by Nguyen et al. presents a case study on updating finite element models of steel bridge structures using vibration data, highlighting the effectiveness of vibration-based monitoring in assessing structural health (Nguyen et al. 2023). The vibration studies should conduct a thorough evaluation, as vibrations may be influenced by the foundation's low strength or the quality of the joiners. The rehabilitation strategies are a crucial aspect of design to ensure a structure that is free from failure.

The study identifies a model method for diagnosing a vibration-affected structure and outlines rehabilitation approaches. By utilizing the capabilities of AI, a similar model can be applied to various structures for the prediction of vibration-related damages. AI-powered methods can also make it easier to keep an eye on the structure while it's being loaded and unloaded, which lets rehabilitation plans be made more quickly.

2 Project Profile

A steel framework supports an extensive arrangement of air fans in the project. The diameter of each air fan was 13 ft, and the structure was designed with a width of 9.0 m, featuring a continuous length to accommodate twenty-eight fans arranged in longitudinal directions. The entire structure was upheld by a concrete pedestal measuring 25 mm in thickness. The elevation of the structure measured 4.5 m from the apex of the fan. The structure features movable motors, a trolley, a ladder, vibration control systems, and catwalks designed for maintenance purposes. The structure supports itself horizontally, with ISMB500 at the top and ISMB200 at the bottom. The vertical members consist of ISMB400. The previous data regarding vibration was identified for the structure. This historical data was analyzed concerning motor drive end vibration in axial directions. The vibration data was collected with high resolutions in the non-directional end of the motor, specifically in the horizontal directions. In a similar manner, the data was collected in vertical orientations. The observations were conducted using motion amplification videos. The preliminary analysis revealed significant vibration and bending in the structure. The second stage of data collection encompassed the diagnosis of the complete structure. The previous data for the project indicated that the machine is producing significant vibrations on the motor and beam structure, with amplitudes recorded at 4.62 and 7.80 mm/s. The beam exhibited bending in vertical directions, as demonstrated by motion amplification videos. The underlying issue could stem from an inadequate supporting structure for the beam, which may be causing elevated vibrations (Figs. 1, 2).

3 Rebound Hammer Test

The rebound hammer test (RHT) was performed on the concrete foundations of the structure. Understanding the effect of vibration on the core part of the structure necessitates identifying the strength of the foundation. The inadequate strength of concrete in the foundation could significantly contribute to the vibrations observed. RHT serves as an effective, non-invasive technique for assessing the strength of concrete. The assessment relies on the rebound characteristics of a spring. Upon impact with the sample, the rebound distance is used to determine the compressive

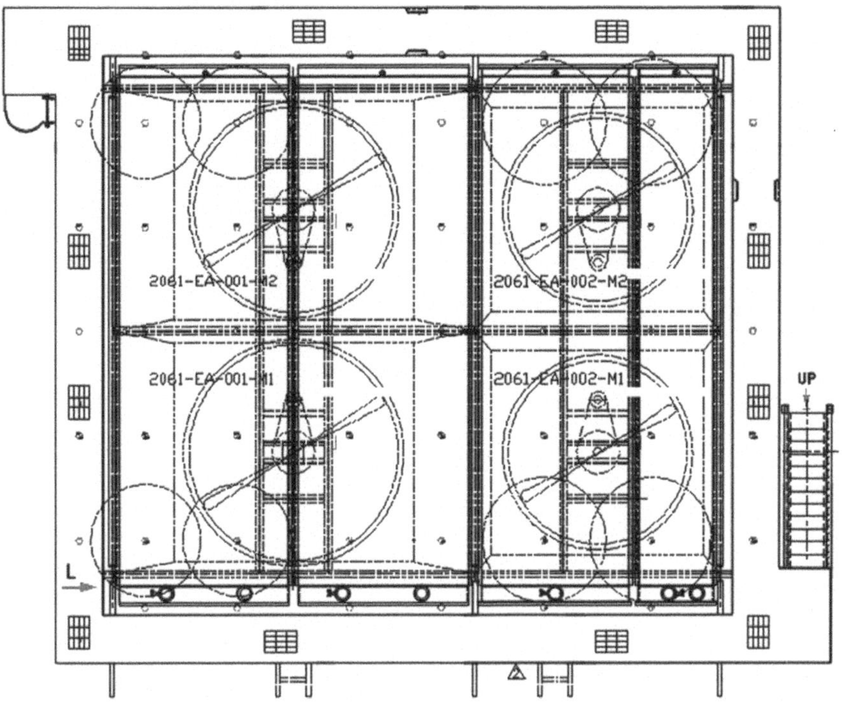

Fig. 1 Air fin structure module

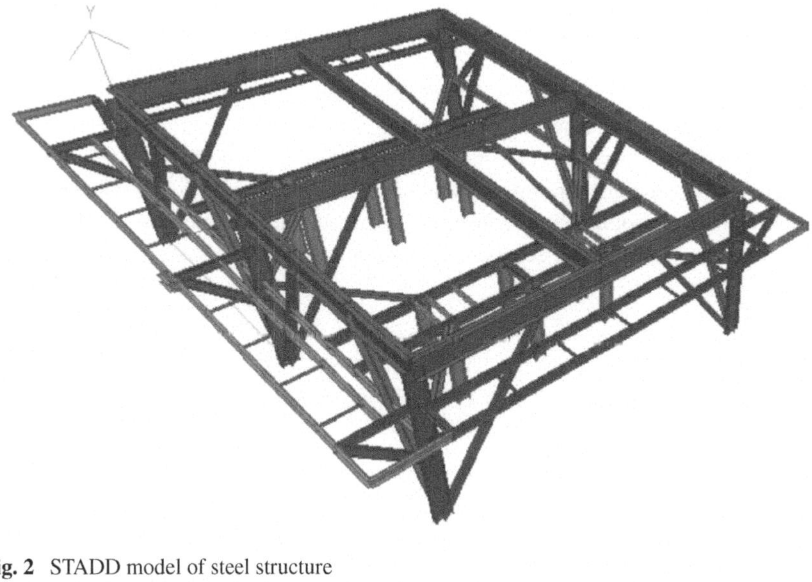

Fig. 2 STADD model of steel structure

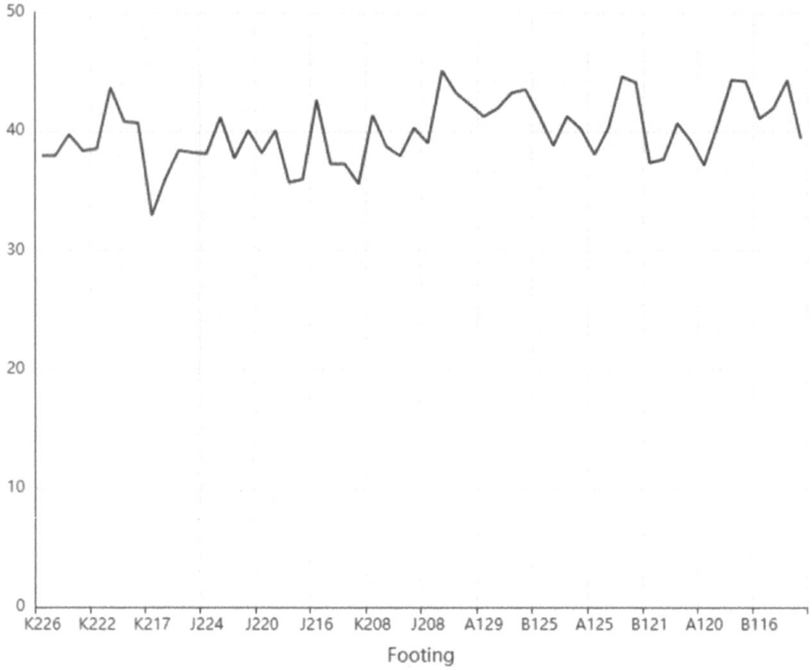

Fig. 3 Results of rebound hammer test

strength of the concrete. The rebound hammer test relies on various factors, including surface characteristics, roughness, hardness, and moisture content. The assessment involves obtaining at least nine measurements from the concrete surface. A rebound index is obtained by averaging nine readings. The rebound index is transformed into the compressive strength of concrete. According to IS: 13311(2) 1992 for non-destructive tests, the compressive strength of concrete must exceed 35 MPa. The findings from the rebound hammer test reveal that the measurements exceed 35 MPa. The findings indicate that the compressive strength meets the necessary requirements (Fig. 3).

4 Vibration Studies

The vibration studies were carried out using a vibration meter that ranges from 0 to 190 m/s^2 and 0 to 199 mm/s. The calibrated meter was employed to gather data on the vibrations of different structural components. The vibration data was incorporated into the STAAD model and analyzed using KNIME. The objective of the STAAD model was to verify the permissible frequency. The STAAD model was utilized to create scenarios representing the conditions "before" and "after". The two scenarios

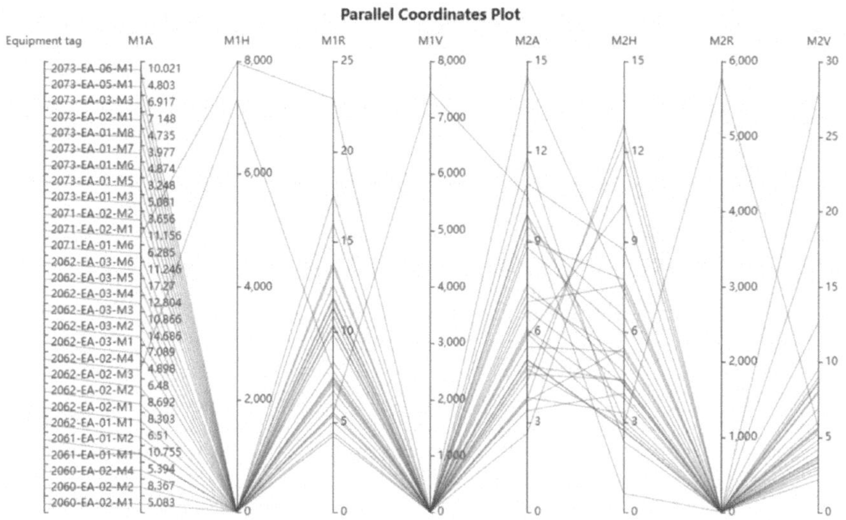

Fig. 4 Vibration old data

were analyzed to assess the rehabilitation method. The KNIME platform facilitated the development of machine learning models utilizing artificial neural networks. The amplitude observed in the columns of the majority structure ranges from 7 to 15.7 mm/s. The range was determined to be exceptionally high. The amplitudes observed in most beams fall within the range of 7–15 mm/s. This range exhibits notable elevation. The frequency in the STAAD analysis is determined to be 12 Hz. Nonetheless, the frequency under the site conditions is exceedingly high. The site studies' observations led to the conclusion that the elevated beam near the fans is impacting them. The vibration analysis indicated that the support beams, directly adjacent to the motor, are experiencing the highest levels of vibration. The primary source of vibration originates from the lighter segment of the beam that directly supports the fan structure. The immediate beam adjacent to the fan has demonstrated significant vertical deflection in the STAAD analysis (Figs. 4, 5).

5 Artificial Neural Network

An artificial neural network is a machine learning model designed to replicate the intricate functions of human brains, similar to the way neurons operate. Artificial Neural Networks consist of interconnected nodes that function similarly to neurons, enabling them to process data effectively. This neuron endeavors to identify patterns, facilitating the processes of pattern recognition and decision-making within a dataset. Artificial Neural Networks present both benefits and drawbacks. The following are

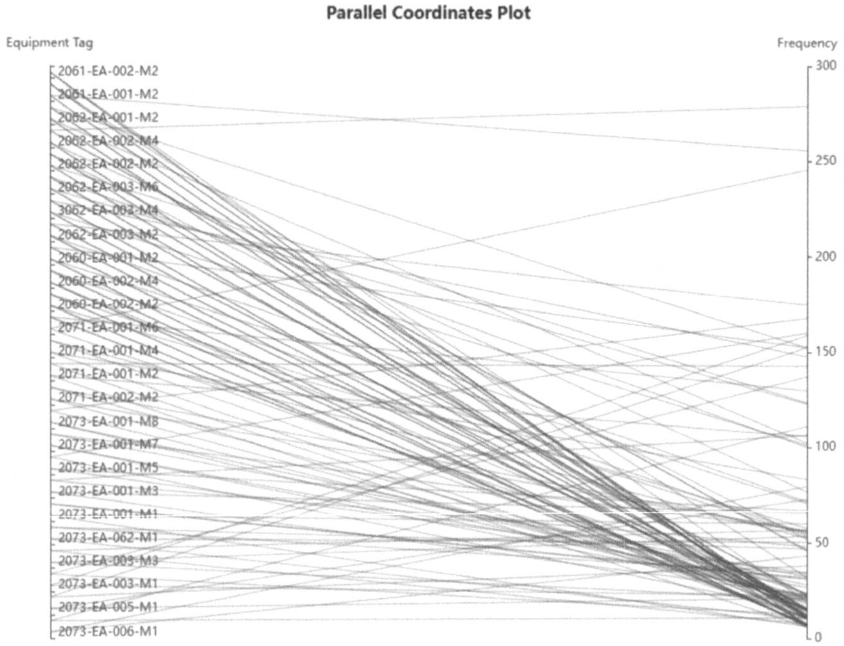

Fig. 5 Vibration new data

some advantages of ANN: (a) It features a flexible structure that adjusts to new circumstances and acquires knowledge from diverse datasets. The relationships between inputs and outputs in vibration studies are intricate, owing to the complexity of the data set involved. (b) ANN exhibits significant strengths in its ability to process complex patterns within data. Tasks involving pattern recognition and image processing can incorporate this. (c) The ability of ANN to process many tasks at once is useful for a vibration dataset because the structure is always changing. The ANN model for vibration data is capable of simultaneously handling multiple processes. (d) Neural networks exhibit nonlinearity, with the model handling data through a non-linear activation function. Nonetheless, neural networks can present certain drawbacks, as they require substantial computing power and larger datasets, potentially impacting performance in vibration datasets. Nonetheless, artificial neural networks have extensive applications beyond vibration analysis, including natural language processing, image and video recognition, healthcare, finance, and numerous other fields. The development of neural networks has its origins in the 1940s, marked by the introduction of a mathematical model for artificial neurons by McCulloch and Pitts. However, throughout this timeframe, there was an insufficiency in computational resources to perform neural network modeling. In the 1960s, Frank Rosenblatt developed perceptions as a straightforward, single-layer network. This addresses a linear issue but proved ineffective for more intricate challenges. In the 1980s, the introduction of backpropagation transformed neural networks by facilitating the training of multi-layer networks. High computational costs and complexity limited the advancement of neural networks during the 1990s. In the year 2000, advancements in deep learning and increased computational power enabled the processing of extensive datasets. Beginning in 2010, the advancement of computational capabilities led to a significant rise in the popularity of neural networks. Neural networks typically structure their architecture in layers, using the next three layers for dataset computation and training. The layers consist of (1) an input layer, (2) hidden layers, and (3) an output layer. The input layer serves as the initial point where the network acquires the data. In these cases, each component of the data set corresponds to the input neural network. The concealed layers constitute the core component of computational tasks. It is possible to have several concealed layers. What happens depends on the features of the data that was used for training. These layers are made up of neurons that process the input data. The output layer ultimately handles particular tasks such as regression and classification, among others. The system landscape is characterized by forward propagation. The processed data transitions from the input layer, passes through the hidden layer, and ultimately reaches the output layer. Neurons in the various layers receive inputs, scaled by the weights associated with their connections, during the initial phase of this process. The sum of these components is calculated, and bias is incorporated into the total. The outcomes of linear transformation undergo an activation function. At this point, the system has established its non-linearity. This enables them to understand intricate patterns. This process is comparable to that of forward propagation. Neural networks examine backward propagation, enabling them to distinguish between the actual output and the predicted output. The objective here is to reduce the loss. Iteration for forward propagation is an important part of the neural network because it makes both training and predicting the dataset more accurate.

6 ANN and KNIME

KNIME is a cutting-edge platform designed for the development of seamless machine learning models, offering scalability and versatility along with robust integration capabilities for various Python libraries. Neural networks in KNIME provide a straightforward method for analyzing and predicting data across a variety of applications. Celik discusses integrating KNIME with an Enterprise Service Bus (ESB) platform to enhance predictive maintenance in industrial settings using ANNs, highlighting improved data transaction and interfacing capabilities (Celik 2018). Tougui et al. compares multiple data mining tools, including KNIME, applying machine learning techniques such as ANNs for heart disease classification, with MATLAB's ANN model achieving the highest performance (Tougui et al. 2020). Abualkibash evaluates various machine learning algorithms, including ANNs, within KNIME for analyzing network security datasets, demonstrating the platform's effectiveness in cybersecurity applications (Abualkibash 2019). Basha et al. assesses the performance

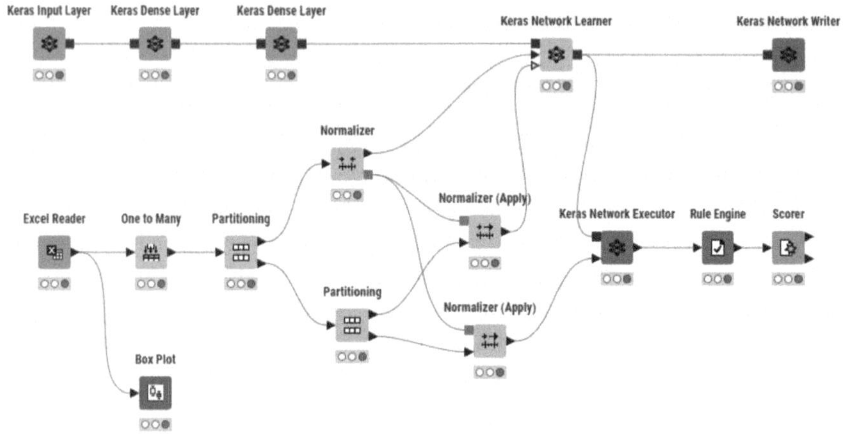

Fig. 6 KNIME work flow

of supervised classification models within KNIME, providing insights into the platform's capabilities for implementing various machine learning algorithms (Basha et al. 2018). Thus, KNIME provides significant capabilities to develop neural networks with code less methods that can be easily integrated with many programming languages.

This project involves the development of a vibration prediction model using workflows in KNIME. The KNIME software is set up for a Python environment to utilize different prediction models. To initiate the development of an ANN in KNIME, the first step involves establishing a Python environment, which can be accomplished through the installation of Anaconda. The establishment of the Conda environment is crucial and involves several key steps. There is a built-in node in KNIME called the Conda Environment Propagation node that takes care of the conda environment needed for neural network processing (Fig. 6).

7 ANN Model

The model has been developed with KNIME Analytical Platform (Fig. 7). The model primarily utilizes Keras nodes. The neural network is constructed with an input layer, hidden layers, and an output layer. The Keras network learner and Keras network executor serve the purpose of training and predicting the dataset. Predictive nodes forecast the data set. A neural network is developed using Python, featuring one hidden layer to process two sets of components: "FAIL" and "PASS". This task can be classified as a multinomial classification task. The initial phase involves the commencement of data processing. The process involves encoding the target. The process involves transforming categories into members, followed by the partitioning of the dataset into training, validation, and test sets. The dataset processes features based on the network and data type. The vibration commercial dataset undergoes processing during the forward propagation phase of the neural network. This is subsequently adjusted to fit within a range of 0–1. The subsequent network architecture is constructed utilizing a Keras input layer. A normalizer node is employed for the input layer to achieve this. This node sets up the input layer in a manner similar to an input tensor. This input tensor functions as a mathematical entity that delineates a multilinear connection among a collection of algebraic objects. To analyze the vibration dataset, a multilinear relationship is established based on the vibration data for training purposes. The relationship of the input shape in the vibration data is characterized by 2 units. Notably, in the KNIME ANN algorithm, it is sufficient to specify the input shape. Additionally, the KNIME ANN algorithm automatically computes the configurations in all remaining layers. The number of input units and the layer arrangement dictate this. The network incorporates a hidden layer using a Keras dense layer node to enhance it and facilitate algebraic relationships between input units and output units. Regarding the vibration data, there are four units. The data is processed into ReLU, which stands for Rectified Linear Units, serving as a mathematical expression to introduce nonlinearity in artificial neural networks. Finally, we establish an additional Keras Dense layer node to serve as the output layer. The vibration data comprises two classes for prediction purposes. Therefore, the output layer is configured to have an output shape of two. Additionally, the ANN is designed to utilize activation functions for a multinomial classification of software functions. This represents a normalized exponential function, transforming a set of real numbers into a probability distribution of potential outcomes. The neural network architecture consists of three layers, featuring a configuration of two, three, and two units in each respective layer. This network architecture includes a Keras network learner, which serves as a node responsible for defining the training parameters and executing the network's training. The Keras network learner consists of three input components. When the model post is made for a certain network, it comes with a data segment for training data and an optional data segment for validating data samples. Following a meticulous process of choosing the input and target data, which enables the target columns to achieve the desired output shape of the network. In the vibration data, two cases will be analyzed, and two additional cases will be evaluated for probability within each class. In this step, we must establish the conversion from integer to number and select an appropriate loss function. The dataset will be processed in a multinomial classification framework, and categorical cross-entropy will be used to judge the results. Ultimately, we establish the parameters for training. The iterations for training hold significant importance, with the training process carried out over 100 epochs. The training was configured with a batch size of 5, and optimization was performed using Adam for effective learning. The Adam optimizer is an adaptive moment estimation algorithm utilized in optimization techniques for gradient descent. The optimizer demonstrates efficiency while utilizing minimal memory resources. The artificial neural network operates within a Conda environment and utilizes propagation to generate the output results. The output logs indicated satisfactory performance (Fig. 8). and the network underwent training on the vibration

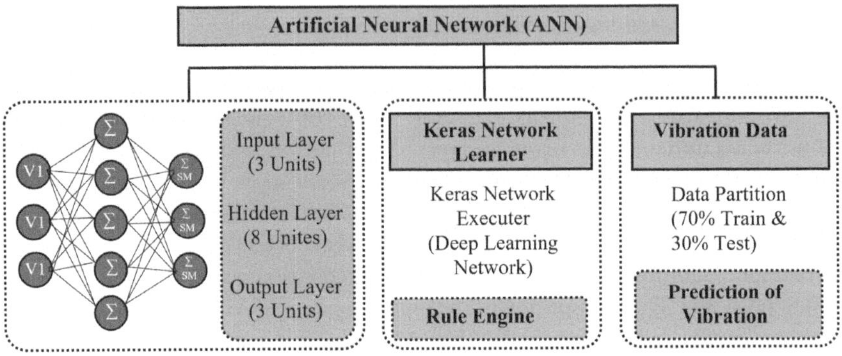

Fig. 7 ANN model

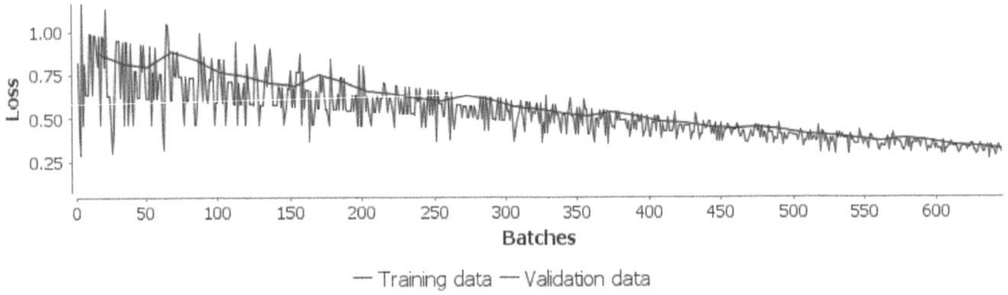

Fig. 8 Network losses

dataset. The network learned from the accumulated knowledge. The Keras Network Executor Node facilitates the prediction process for the vibration dataset. The node generates two supplementary columns derived from the likelihood of each class. The rule engine node determines the maximum probability for the ultimate prediction. The scorer ultimately evaluated the classification model's performance. The Keras network writer node facilitates the saving of the network, enabling its application for predicting datasets (Figs. 7, 8).

8 Rehabilitation of Structure

STAAD models provided deeper insights into the vibration, leading to the rehabilitation of the steel structure. The STAAD models were developed for two scenarios: (1) Before condition: Current loading and vibration conditions, and (2) After condition: Post-retrofitting loading and vibration conditions. The STAAD provides numerous advanced features for modeling steel structures within the software. This facilitates a thorough examination of the structure's behavior for both analysis and design purposes. The finite element method is typically employed to simulate structural responses under dynamic loads. The analysis was conducted under dynamic loading

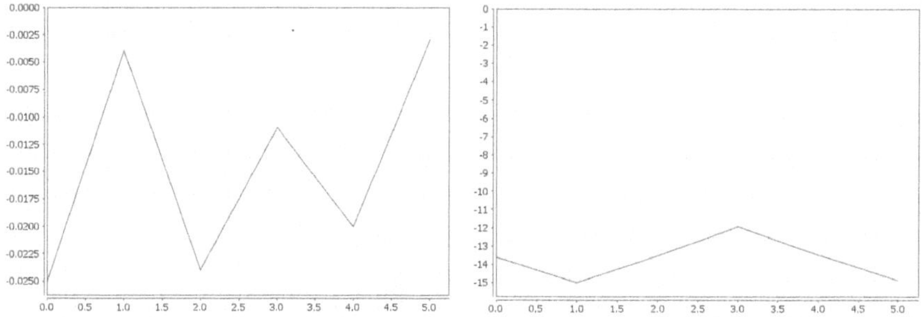

Fig. 9 Deflection in member before and after

conditions to simulate harmonic forces, as well as live and dead loads. The analysis was conducted under dynamic loading conditions to simulate live loads, dead loads, seismic loads, wind loads, and loads resulting from a rotating motor. The calculations of natural frequency were performed for both the pre- and post-conditions of the structure. The Y-axis deflection for the affected beam in the before condition was a maximum of –15.00 mm. In the subsequent analysis, the deflections were recorded at –0.004 mm (Fig. 9). The outlined retrofitting was executed in the following manner:

(a) Bracing support: Two ISA110 x 110 x 10 were installed between the columns and the affected beam for bracing support.
(b) Vertical prop support: Two ISMC300 vertical prop supports were installed on the affected beam to mitigate vibration issues.
(c) Beam support: The beam support was implemented in ISMB400 to mitigate the transmission of vibration.
(d) Column-Beam bracket connection: A 20 mm thick connection between the column and beam was proposed to mitigate vibration issues. The proposed bracket was present at every column and beam throughout the entire structure (Figs. 9, 10, 11).

9 Application of AI for Infrastructure Rehabilitation

AI-based applications for fixing up infrastructure could completely change cities and industries with their advanced ways of researching, analyzing, and guessing what will happen. Across the globe, the infrastructure in cities continues to age. Furthermore, the industrial infrastructure requires consistent upkeep and oversight as it ages. The maintenance of aging infrastructure, along with structural and human safety, as well as the sustainability of industrial facilities, presents significant concerns. The significance of timely infrastructure rehabilitation lies in its ability to save costs and prevent critical damages to industrial operations. The infrastructure necessitates

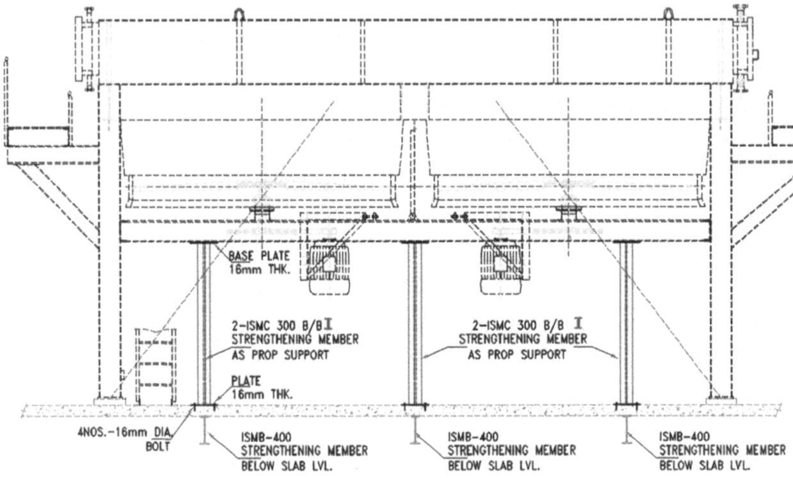

Fig. 10 Proposed strengthening of structure

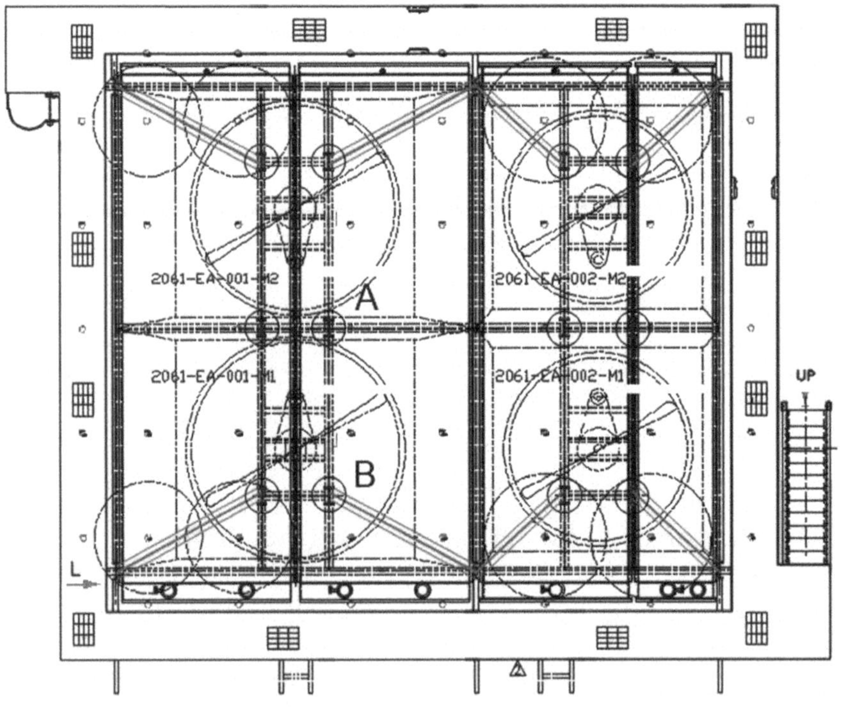

Fig. 11 Proposed strengthening of structure (A—vertical support and B—inclined bracing support)

repairs, restoration, and upgrades, along with regular maintenance, due to deterioration from aging. Zang et al. has talked about deep learning models outperform traditional methods in diagnosing faults in rotating machinery using vibration signals (Zhang et al. 2019). Further Azimi et al. has indicated that AI methods enhance the accuracy of vibration-based damage detection in structures by automating feature extraction and decision-making (Azimi et al. 2020). Worden et al. also has indicated that machine learning models can provide accurate predictions for structural health monitoring based on vibration data (Worden and Hensman 2017). Combining vibration signals with machine learning leads to effective bearing fault detection and classification (Hoang and Kang 2019). Wen et al. has further looked into deep learning can automatically extract features from vibration signals, reducing reliance on manual feature engineering (Wen et al. 2017). Even, Yuen et al. has also indicated that Neural networks accurately localize structural damage by learning spatial patterns in vibration data (Yuen and Kuok 2010). Moreover, AI-driven damage prognosis systems predict remaining useful life and failure risks based on vibration analysis (Wang 2020).

Thus, predictive maintenance of infrastructure is a major application of artificial intelligence with models of machine learning to predict components and system which is likely to fail. As per Chen et al., AI models improve system identification accuracy in civil structural health monitoring applications (Chen and Spencer 2021). This facilitates timely intervention to prevent any potential harm and threats to life. Moreover, AI facilitates the proactive planning of rehabilitation strategies to prevent any potential damage to the structure. Nonetheless, numerous obstacles exist in the application of AI for the rehabilitation of infrastructure. Despite numerous qualities, the key issue that persists is data quality. High-quality data that is accurate is essential for developing effective AI models and machine learning systems. The machine learning models must process the data on a substantial scale.

This paper proposes a neural network application to periodically predict data from vibration meters and sensors. This study introduces a neural network that can regularly gather and utilize such data.

10 Conclusions

A neural network model made to predict vibration data using Python and the Keras framework is shown in an application that focuses on fixing up infrastructure that has been damaged by vibration. The developed neural network consists of an input layer, a hidden layer, and an output layer designed for data prediction. Similar models can be used on many other vibration datasets, including those for bridges, roads, railways, and more, to figure out what the effects of vibrations will be and how they will happen. Additionally, hybrid models can advance the structural design and formulate rehabilitation strategies. Key components of the complete process include the availability of data sets and the development of networks. Vibration sensors typically gather vibration data. The structure can integrate these sensors to continuously gather and analyze data, thereby facilitating a deeper understanding of the structural issues. The subsequent points outline the significant benefits and drawbacks of utilizing AI in the rehabilitation process.

(1) The collection of the data set must be ongoing and should consist of a substantial amount of data to facilitate training and enable accurate predictions. The rehabilitation of the above structure has faced limitations due to insufficient data. However, the model focused on enhancing its capabilities to process data for ongoing analysis and prediction.
(2) One significant benefit of utilizing AI is its ability to forecast structural failures in a brief period. This prediction can prevent any structural failures in the future, thereby saving significant costs in management and maintenance. Because of this, it is important to look at the model using similar neural networks when talking about infrastructure that is getting old in cities and factories. This could save a lot of money on maintenance costs.
(3) The ANN serves as a straightforward model for implementation in intricate tasks within urban environments and industrial settings. Often, due to time limitations, it is not feasible to create a costly predictive model. The process of collecting data through vibration can be quite time-consuming and costly. So, the model shows that it is possible to turn a simple set of vibrations collected by a vibration meter into a strong neural network model.
(4) To formulate effective strategies for the rehabilitation of structures, it is crucial to integrate these models with structural analysis and design techniques. However, the model has constraints when it comes to predicting data, making it challenging to utilize for determining appropriate rehabilitation strategies.

So, combining the neural network model with the right techniques for structural analysis and design can lead to ways to fix the vibrations that are hurting the structure. Comparable models can be created for different types of infrastructure, such as bridges, flyovers, roads, and more, to forecast structural failure.

Declarations

- Author contribution: Parlewar P. contributed to the design and implementation of the research, to the analysis of the results and to the writing of the manuscript.
- Competing interests: There are no competing interests related to this article.
- Funding: Not applicable
- Availability of data and materials: Not applicable

References

Abualkibash M (2019) Machine learning in network security using KNIME analytics. Int J Netw Sec Applic 11(5):1–14

Almutairi KM, Sinha JK (2023) Experimental vibration data in fault diagnosis: a machine learning approach to robust classification of rotor and bearing defects in rotating machines. Machines 11(10):943

Azimi M, Eslamlou AD, Pekcan G (2020) A review on artificial intelligence methodologies for vibration-based damage detection in structures. Sensors 20(18):5233. https://doi.org/10.3390/s20185233

Basha SM, Rajput DS, Poluru RK, Bhushan SB, Basha SAK (2018) Evaluating the performance of supervised classification models: decision tree and Naïve Bayes using KNIME. Int J Eng Technol 7(4.5):248–253

Brzyski J (2019) Assessment of vibration effects on buildings and their rehabilitation. J Civ Eng Manag 25(3):229–240

Celik H (2018) ESB platform integrating KNIME data mining tool oriented on Industry 4.0 based on artificial neural network predictive maintenance. Int J Artif Intell Appl 9(3):1–12

Chen HY, Lee CH (2021) Deep learning approach for vibration signals applications. Sensors 21(11):3929

Chen C, Spencer BF (2021) AI-based system identification for vibration monitoring in civil structures. Eng Struct 243:112603. https://doi.org/10.1016/j.engstruct.2021.112603

Chen J, Yang L, Huang Q, Zhang W, Qiao J, Zhang X, Jiang D (2024) Dynamic analysis of drill string vibration enhanced by machine learning models. SSRN Electron J

Costa TB, Soares JLL, Silva JC, Sousa WS, Mesquita ALA, Mesquita ALA, Figueiredo JMS, Braga DS (2024) Vibration data-driven supervised learning for low-speed bearing fault diagnosis using classical and nonlinear features. SSRN Electron J

Hakim SJS, Abdul Razak H (2011) Application of artificial neural network on vibration test data for damage identification in bridge girder. Int J Phys Sci 6(35):7991–8001

Hoang DT, Kang HJ (2019) Condition monitoring of bearings using vibration signals and machine learning. Sensors 19(9):2062. https://doi.org/10.3390/s19092062

Khan MA, Akhtar K, Ahmad N, Shah F, Khattak N (2020) Vibration analysis of damaged and undamaged steel structure systems: cantilever column and frame. Earthq Eng Eng Vib 19:725–737

Kuniyoshi F, Sawada Y (2024) Can transformers predict vibrations? https://arxiv.org/abs/2402.10511

Lakade G (2019) Analysis and design of steel frame structure for rotating motors. Int J Eng Res Technol 8(8):640–645

Lee XY, Kumar A, Vidyaratne L, Rao AR, Farahat A, Gupta C (2023) An ensemble of convolution-based methods for fault detection using vibration signals

Li H, Wang T, Wu G (2022) A Bayesian deep learning approach for random vibration analysis of bridges subjected to vehicle dynamic interaction. Mech Syst Signal Process 170:108799

Mey O, Neudeck W, Schneider A, Enge-Rosenblatt O (2020) Machine learning-based unbalance detection of a rotating shaft using vibration data

Mohammed HA, Ahmed F (2021) Rehabilitation of bridge structures subjected to dynamic loads. J Bridge Eng 26(7), 04021014

Nguyen DC, Salamak M, Katunin A, Poprawa G (2023) Finite element model updating of steel bridge structure using vibration-based structural health monitoring system: a case study of railway steel arch bridge in Poland. Experiment Vibr Anal Civil Eng Struct, pp 371–380

Papanikolaou V, Kartsaklis V (2020) Assessment and rehabilitation methods for structures subjected to vibrations. Eng Struct 218:110752

Pham HC, Yang Z (2022) Prediction of Bolt loosening using vibrational analysis and machine learning. IOSR J Mech Civil Eng 20(1):27–36

Sharma JK, Sharma S (2024) Application of artificial neural network models for condition monitoring of industrial fan. Int J Intell Syst Appl Eng 11(10s):574–590. Retrieved from https://www.ijisae.org/index.php/IJISAE/article/view/3312

Tougui I, Jilbab A, El Mhamdi J (2020) Heart disease classification using data mining tools and machine learning techniques. Heal Technol 10:1137–1144

Vallim A, de Souza GFM (2019) Prediction of motor failure time using an artificial neural network. Sensors 19(19):4342

Van Delden J, Schultz J, Blech C, Langer SC, Lüddecke T (2023) Learning to predict structural vibrations. arXiv preprint

Wang D, Chen X (2020) AI-based damage prognosis using vibration data. J Sound Vibr 470:115175. https://doi.org/10.1016/j.jsv.2020.115175

Wen L, Li X, Gao L, Zhang Y (2017) Application of deep learning to fault detection in rotating machinery. Neurocomputing 237:227–237. https://doi.org/10.1016/j.neucom.2017.01.045

Worden K, Hensman JJ (2017) Data-driven vibration-based structural health monitoring using machine learning. Philos Trans Royal Soc A 375(2096):20160266. https://doi.org/10.1098/rsta.2016.0266

Yuen KV, Kuok SC (2010) Neural networks for vibration-based structural damage localization. J Sound Vibr 329(24):5160–5176. https://doi.org/10.1016/j.jsv.2010.06.021

Zhang W, Peng G, Li C, Chen Y (2019) Deep learning for fault diagnosis of rotating machinery: a review. Mech Syst Signal Process 108:33–47. https://doi.org/10.1016/j.ymssp.2018.02.016

The manufacturer's authorised representative in the EU is Springer Nature Customer Service Centre GmbH, Europaplatz 3, 69115 Heidelberg, Germany. If you have any concerns regarding our products, please contact ProductSafety@springernature.com

Printed and bound by CPI Group (UK) Ltd, Croydon, CR0 4YY
07/10/2026
02229361-0006